SpringerBriefs present concise summaries of cutting-edge research and practical applications across a wide spectrum of fields. Featuring compact volumes of 50 to 125 pages, the series covers a range of content from professional to academic.

Typical topics might include:

- A timely report of state-of-the art analytical techniques
- A bridge between new research results, as published in journal articles, and a contextual literature review
- A snapshot of a hot or emerging topic
- An in-depth case study or clinical example
- A presentation of core concepts that students must understand in order to make independent contributions

Briefs allow authors to present their ideas and readers to absorb them with minimal time investment. Briefs will be published as part of Springer's eBook collection, with millions of users worldwide. In addition, Briefs will be available for individual print and electronic purchase. Briefs are characterized by fast, global electronic dissemination, standard publishing contracts, easy-to-use manuscript preparation and formatting guidelines, and expedited production schedules. We aim for publication 8–12 weeks after acceptance. Both solicited and unsolicited manuscripts are considered for publication in this series.
**Indexing: This series is indexed in Scopus and zbMATH **

More information about this series at http://www.springer.com/series/10028

SpringerBriefs in Computer Science

Xiaojun Yuan • Zhipeng Xue

Turbo Message Passing Algorithms for Structured Signal Recovery

Springer

Xiaojun Yuan
Center for Intelligent Networking & Communication
University of Electronic Science & Technology of China
Chengdu, China

Zhipeng Xue
School of Information Science and Technology
ShanghaiTech University
Shanghai, China

ISSN 2191-5768 ISSN 2191-5776 (electronic)
SpringerBriefs in Computer Science
ISBN 978-3-030-54761-5 ISBN 978-3-030-54762-2 (eBook)
https://doi.org/10.1007/978-3-030-54762-2

This Springer imprint is published by the registered company Springer Nature Switzerland AG
The registered company address is: Gewerbestrasse 11, 6330 Cham, Switzerland

Preface

Message passing is an efficient iterative algorithm for solving inference, optimization, and satisfaction problems. It has found successful applications in numerous areas, including channel coding (e.g., low-density parity check (LDPC) codes and turbo codes), computer vision (e.g., stereo images), and free energy approximation. Recently, novel message passing algorithms, such as approximate message passing (AMP) and turbo compressed sensing (Turbo-CS), have been developed for sparse signal reconstruction. While the AMP algorithm exhibits guaranteed performance for sensing matrices with independent and identically distributed (i.i.d.) elements, Turbo-CS achieves a much better performance when the sensing matrix is constructed based on fast orthogonal transforms (such as the discrete Fourier transform and the discrete cosine transform). Inspired by the success of Turbo-CS, a series of message passing algorithms have been developed for solving various structured signal recovery problems with compressed measurements. We call these algorithms turbo message passing algorithms. In this book, we will undertake a comprehensive study on turbo message passing algorithms for structured signal recovery, where the structured signals include: (1) a sparse vector/matrix (which corresponds to the compressed sensing (CS) problem), (2) a low-rank matrix (which corresponds to the affine rank minimization (ARM) problem), and (3) a mixture of a sparse matrix and a low-rank matrix (which corresponds to the robust principal component analysis (RPCA) problem). In particular, the book is divided into the following parts: First, for the CS problem, we introduce a turbo message passing algorithm termed denoising-based Turbo-CS (D-Turbo-CS). We show that with D-Turbo-CS, signals without the knowledge of the prior distributions, such as images, can be recovered from compressed measurements by using message passing with a very low measurement rate. Second, we introduce a turbo message passing (TMP) algorithm for solving the ARM problem. We further discuss the impact of the use of various linear operators on the recovery performance, such as right-orthogonally invariant linear (ROIL) operators and random selectors. Third, we introduce a TMP algorithm for solving the compressive RPCA problem which aims to recover a low-rank matrix and a sparse matrix from their compressed mixture. We then apply the TMP algorithm to the video background subtraction problem and show that

TMP achieves much better numerical and visual recovery performance than its counterparts. With this book, we wish to spur new researches on applying message passing to various inference problems.

Chengdu, China
Shanghai, China
June 2020

Xiaojun Yuan
Zhipeng Xue

Acknowledgements

This book has become a reality with the kind support and help of many individuals in the past few years. We would like to extend our sincere thanks to all of them.

Contents

Chapter 1
Introduction

1.1 Background

With the rapid development of the Internet and Internet of Things (IoT) technologies, huge amounts of data are generated by personal mobile devices, intelligent IoT devices, and sensors. The collection, transmission, and storage of large amounts of data become a crucial problem, and reducing data sizes by exploiting data structures becomes an essential technology for many applications.

Compressed sensing (CS) is a new sampling paradigm that exploits the sparsity of signals. The basic idea is that in many applications a sparse representation of signals can be found on a chosen basis. Besides sparse representation, many real-world signals such as voice signals, images, and videos contain information-redundant structures and so can be compressed without much information loss. Below is a list of structured signals to be considered in this book:

- sparse signals (voice, wireless array channels),
- low-rank signal (images, data in recommendation systems),
- the mixture of a sparse signal and a low-rank signal (system identification, surveillance video).

The recovery of structured data from compressed measurements can be cast as an inference problem with the prior knowledge of data structure. Message passing is an efficient inference method that has found wide applications in many areas including optimization, inference, and constraint satisfaction. The basic idea of message passing is to pass messages between factor nodes and variable nodes on a factor graph until the messages converge. It has proved to be exact on tree-type graphical models and can also be applied to graphical models with loops.

Recently, message passing has been applied to sparse signal recovery, which leads to low-complexity and fast-convergence algorithms such as the approximate message passing algorithm (AMP) [1] for CS with independent and identically

X. Yuan and Z. Xue, *Turbo Message Passing Algorithms for Structured Signal Recovery*, SpringerBriefs in Computer Science,
https://doi.org/10.1007/978-3-030-54762-2_1

distributed (i.i.d.) Gaussian sensing matrix and the turbo compressed sensing algorithm [2] for CS with partial orthogonal sensing matrix. In this book, we start with a general description of the turbo message passing principle for structured signal recovery. Next, we briefly introduce the design of the turbo message passing algorithm for the recovery of three types of structured signals: the compressed sensing problem [3], the low-rank matrix recovery problem [4], and the compressed robust principal component analysis problem [5].

1.1.1 Compressed Sensing

Compressed sensing is a new sampling paradigm that exploits the sparsity of signals. A common approach for compressed sensing is to solve a mixed l_1-norm and l_2-norm minimization problem via convex programming [6]. However, a convex program in general involves polynomial-time complexity, which causes a serious scalability problem for mass data applications.

Approximate algorithms have been extensively studied to reduce the computational complexity of sparse signal recovery. Existing approaches include match pursuit [7], orthogonal match pursuit [8], iterative soft thresholding [9], compressive sampling matching pursuit [10], and approximate message passing [1]. In particular, AMP is a fast-convergence iterative algorithm based on the principle of message passing. It has been shown that when the sensing matrix is i.i.d. Gaussian, AMP is asymptotically optimal as the dimension of the state space goes to infinity [11]. Also, the iterative process of AMP can be tracked through a scalar recursion called state evolution.

In many applications, compressive measurements are taken from a transformed domain, such as discrete Fourier transform (DFT), discrete cosine transform (DCT), and wavelet transform, etc. This, on the one hand, can exempt us from storing the sensing matrix in implementation; on the other hand, these orthogonal transforms can be realized using fast algorithms to reduce the computational complexity. However, the AMP algorithm, when applied to orthogonal sensing, does not perform well and its simulated performance deviates away from the prediction by the state evolution.

Turbo compressed sensing (Turbo-CS) [2] solves the above discrepancy by a careful redesign of the message passing algorithm. The Turbo-CS algorithm consists of two processing modules: One module handles the linear measurements of the sparse signal based on the linear minimum mean square error (LMMSE) principle and calculates the so-called extrinsic information to decorrelate the input and output estimation errors; the other module combines its input with the signal sparsity by following the minimum mean square error (MMSE) principle and also calculates the extrinsic information. The two modules are executed iteratively to refine the estimates. This is similar to the decoding process of a turbo code [12], hence the name Turbo-CS. It has been shown that Turbo-CS considerably outperforms its counterparts for compressed sensing in both complexity and convergence speed.

1.1.2 Low-Rank Matrix Recovery

Low-rank matrices have found extensive applications in real-world applications including but not limited to remote sensing [13], recommendation systems [14], global positioning [15], and system identification [16]. In these applications, a fundamental problem is to recover an unknown matrix from a small number of observations by exploiting its low-rank property [4, 17]. Specifically, we want to recover a low-rank matrix $\boldsymbol{X}_0$ from the compressed measurement $\boldsymbol{y} = \mathcal{A}(\boldsymbol{X}_0)$. A special case of the low-rank matrix recovery problem is the low-rank matrix completion problem where $\mathcal{A}$ is a random selector. The low-rank matrix recovery problem can be cast as an affine rank minimization (ARM) problem. However, this problem is NP-hard, and so solving ARM is computationally prohibitive. To reduce complexity, a popular alternative to ARM is to solve the nuclear norm minimization (NNM) problem. In [18], Recht et al. proved that when the restricted isometry property (RIP) holds for the linear operator $\mathcal{A}$, the ARM problem is equivalent to the NNM problem. The NNM problem can be solved by semidefinite programming (SDP). Existing convex solvers, such as the interior point method [16], can be employed to find a solution in polynomial time. However, SDP is computationally heavy, especially when applied to large-scale problems with high-dimensional data. To address this issue, low-cost iterative methods, such as the singular value thresholding (SVT) method [19] and the proximal gradient algorithm [20], have been proposed to further reduce complexity at the cost of a certain amount of performance degradation.

1.1.3 Compressed Robust Principal Component Analysis

As a modified statistical principal component analysis method, robust principal component analysis (RPCA) aims to recover a low-rank matrix $\boldsymbol{L}$ from highly corrupted measurements of $\boldsymbol{L} + \boldsymbol{S}$, where $\boldsymbol{S}$ is a sparse matrix that represents the grossly corrupted entries. RPCA finds a wide range of applications in video surveillance [5], face recognition [21], and information retrieval [22], etc. In the literature, there are a number of different RPCA approaches for the decomposition of low-rank and sparse matrices, including principal component pursuit (PCP) [5, 23], the iterative thresholding (IT) method [24], the accelerated proximal gradient (APG) method [25], the augmented Lagrange multiplier (ALM) method [26], and the alternating direction method (ADM) [27].

As inspired by the success of RPCA, people are curious about to what extent the linear measurements of $\boldsymbol{L} + \boldsymbol{S}$ can be compressed without comprising the capability to recover low-rank matrix $\boldsymbol{L}$ and sparse matrix $\boldsymbol{S}$. This gives rise to the problem of compressed RPCA that is useful for the development of new sensing architectures in many applications. Existing methods for RPCA, such as the IT and APG methods, can be modified to solve compressed RPCA. Further developments on compressed

RPCA include [28–31]. In [28], a greedy algorithm, termed SpaRCS, was proposed to iteratively estimate the low-rank matrix by using the ADMiRA method in [32] and the sparse matrix by using the CoSaMP method in [10]. It was reported that SpaRCS achieves better performance than the IT and APG methods. In [29, 33], the authors proposed a variant of PCP and established exact recovery conditions for noise-free compressed RPCA. To handle the measurement noise, the stable PCP (SPCP) method is proposed in [30]. In [31], the authors developed an alternative regularized optimization algorithm to solve the compressed RPCA problem with provable performance in the presence of measurement noise.

1.2 Contributions

In this book, we focus on the algorithm design for the recovery of the structural signals based on the turbo message passing principle. The key contributions of this book are summarized in the following aspects.

- The turbo message passing principle is introduced for structured signal recovery.
- Sparse signal recovery and compressive image recovery: Based on the turbo principle, we extend the turbo compressed sensing algorithm to handle general sparsity and image structures, and establish the state evolution analysis.
- Low-rank matrix recovery: We present a message passing algorithm for both low-rank matrix recovery and low-rank matrix completion based on the turbo principle. The state evolution analysis is established for the low-rank matrix recovery problem when the linear operator is chosen as a right-orthogonally invariant linear operator.
- Compressed robust principal component analysis: A turbo-type algorithm framework is proposed for compressed robust principal component analysis. The proposed algorithm achieves a faster convergence rate than the counterpart algorithms.

1.3 Organization

The book is organized as follows. Chapter 2 describes the turbo message passing principle based on which algorithms are derived for various structured signals. Chapter 2 also describes the state evolution of the turbo message passing algorithm when partial orthogonal sensing matrices are involved. Chapter 3 proposes an algorithm for low-rank matrix recovery and matrix completion problems by following the turbo message passing principle, and the state evolution analysis is established for the right-orthogonally invariant linear operators. In Chap. 4, a turbo message passing framework is established for compressed robust principal component analysis problem, together with the applications to the video background subtraction.

References

1. D.L. Donoho, A. Maleki, A. Montanari, Message-passing algorithms for compressed sensing. Proc. Natl. Acad. Sci. USA **106**(45), 18914–18919 (2009)
2. J. Ma, X. Yuan, L. Ping, Turbo compressed sensing with partial DFT sensing matrix. IEEE Signal Process. Lett. **22**(2), 158–161 (2015)
3. D.L. Donoho, Compressed sensing. IEEE Trans. Inf. Theory **52**(4), 1289–1306 (2006)
4. E.J. Candes, Y. Plan, Matrix completion with noise. Proc. IEEE **98**(6), 925–936 (2010)
5. E.J. Candès, X. Li, Y. Ma, J. Wright, Robust principal component analysis? J. ACM **58**(3), 11 (2011)
6. R. Tibshirani, Regression shrinkage and selection via the lasso. J. R. Stat. Soc. Series B Stat. Methodol. **58**, 267–288 (1996)
7. S.G. Mallat, Z. Zhang, Matching pursuits with time-frequency dictionaries. IEEE Trans. Signal Process. **41**(12), 3397–3415 (1993)
8. J.A. Tropp, Greed is good: algorithmic results for sparse approximation. IEEE Trans. Inf. Theory **50**(10), 2231–2242 (2004)
9. R.D. Nowak, S.J. Wright et al., Gradient projection for sparse reconstruction: application to compressed sensing and other inverse problems. IEEE J. Sel. Top. Signal Process. **1**(4), 586–597 (2007)
10. D. Needell, J.A. Tropp, CoSaMP: Iterative signal recovery from incomplete and inaccurate samples. Appl. Comput. Harmon. Anal. **26**(3), 301–321 (2009)
11. M. Bayati, A. Montanari, The dynamics of message passing on dense graphs, with applications to compressed sensing. IEEE Trans. Inf. Theory **57**(2), 764–785 (2011)
12. C. Berrou, A. Glavieux, Near optimum error correcting coding and decoding: turbo-codes. IEEE Trans. Commun. **44**(10), 1261–1271 (1996)
13. R. Schmidt, Multiple emitter location and signal parameter estimation. IEEE Trans. Antennas Propag. **34**(3), 276–280 (1986)
14. D. Goldberg, D. Nichols, B.M. Oki, D. Terry, Using collaborative filtering to weave an information tapestry. Commun. ACM **35**(12), 61–70 (1992)
15. A. Singer, A remark on global positioning from local distances. Proc. Natl. Acad. Sci. USA **105**(28), 9507–9511 (2008)
16. Z. Liu, L. Vandenberghe, Interior-point method for nuclear norm approximation with application to system identification. SIAM J. Matrix Anal. Appl. **31**(3), 1235–1256 (2009)
17. M.A. Davenport, J. Romberg, An overview of low-rank matrix recovery from incomplete observations. IEEE J. Sel. Top. Signal Process. **10**(4), 608–622 (2016)
18. B. Recht, M. Fazel, P.A. Parrilo, Guaranteed minimum-rank solutions of linear matrix equations via nuclear norm minimization. SIAM Rev. **52**(3), 471–501 (2010)
19. J.-F. Cai, E.J. Candès, Z. Shen, A singular value thresholding algorithm for matrix completion. SIAM J. Optim. **20**(4), 1956–1982 (2010)
20. K.-C. Toh, S. Yun, An accelerated proximal gradient algorithm for nuclear norm regularized linear least squares problems. Pac. J. Optim. **6**, 615–640, 15 (2010)
21. J. Wright, A. Ganesh, S. Rao, Y. Peng, Y. Ma, Robust principal component analysis: exact recovery of corrupted low-rank matrices via convex optimization, in *Proc. of Advances in Neural Information Processing Systems (NeurIPS)*, Vancouver, June 2009, pp. 2080–2088
22. C.H. Papadimitriou, P. Raghavan, H. Tamaki, S. Vempala, Latent semantic indexing: a probabilistic analysis. J. Comput. Syst. Sci. **61**(2), 217–235 (2000)

23. Z. Zhou, X. Li, J. Wright, E. Candes, Y. Ma, Stable principal component pursuit, in *Proc. of IEEE International Symposium on Information Theory (ISIT)*, Austin, June 2010, pp. 1518–1522
24. J. Wright, A.Y. Yang, A. Ganesh, S.S. Sastry, Y. Ma, Robust face recognition via sparse representation. IEEE Trans. Pattern Anal. Mach. Intell. **31**(2), 210–227 (2008)
25. M. Chen, A. Ganesh, Z. Lin, Y. Ma, J. Wright, L. Wu, Fast convex optimization algorithms for exact recovery of a corrupted low-rank matrix. Coordinated Science Laboratory Report no. UILU-ENG-09-2214, DC-246 (2009)
26. Z. Lin, M. Chen, Y. Ma, The augmented Lagrange multiplier method for exact recovery of corrupted low-rank matrices. https://arxiv.org/abs/1009.5055
27. X. Yuan, J. Yang, Sparse and low-rank matrix decomposition via alternating direction methods. Pac. J. Optim. **9**(1), 167–180 (2013)
28. A.E. Waters, A.C. Sankaranarayanan, R. Baraniuk, SpaRCS: recovering low-rank and sparse matrices from compressive measurements, in *Proc. of Advances in Neural Information Processing Systems (NeurIPS)*, Granada, Dec 2011, pp. 1089–1097
29. A. Ganesh, K. Min, J. Wright, Y. Ma, Principal component pursuit with reduced linear measurements, in *Proc. of IEEE International Symposium on Information Theory (ISIT)*, Cambridge, July 2012, pp. 1281–1285
30. A. Aravkin, S. Becker, V. Cevher, P. Olsen, A variational approach to stable principal component pursuit, in *Proc. of Conference on Uncertainty in Artificial Intelligence (UAI)*, Quebec, July 2014, pp. 32–41
31. W. Ha, R.F. Barber, Robust PCA with compressed data, in *Proc. of Advances in Neural Information Processing Systems (NeurIPS)*, Montreal, Dec 2015, pp. 1936–1944
32. K. Lee, Y. Bresler, ADMiRA: atomic decomposition for minimum rank approximation. IEEE Trans. Inf. Theory **56**(9), 4402–4416 (2010)
33. J. Wright, A. Ganesh, K. Min, Y. Ma, Compressive principal component pursuit. Inf. Inference J. IMA **2**(1), 32–68 (2013)

Chapter 2
Turbo Message Passing for Compressed Sensing

2.1 Preliminaries

2.1.1 Compressed Sensing

Consider the following real-valued linear system:

$$y = Ax + n \tag{2.1}$$

where $x \in \mathbb{R}^n$ is an unknown signal vector, $A \in \mathbb{R}^{m \times n}$ is a known constant matrix, and n is a white Gaussian noise vector with zero mean and covariance $\sigma^2 I$. Here, I represents the identity matrix of an appropriate size. Our goal is to recover x from the measurement y. In particular, this problem is known as compressed sensing when $m \ll n$ and x is sparse.

Basis Pursuit De-Noising (BPDN) is a well-known approach to the recovery of x in compressed sensing, with the problem formulated as

$$\hat{x} = \arg\min_{x \in \mathbb{R}^N} \frac{1}{2}\|y - Ax\|_2^2 + \lambda\|x\|_1. \tag{2.2}$$

where $\|x\|_p = (\sum_n |x_n|^p)^{1/p}$ represents the l_p-norm, x_i is the i-th entry of x, and λ is a regularization parameter. This problem can be solved by convex programming algorithms, such as the interior point method [1] and the proximal method [2]. Interior point method has cubic computational complexity, which is too expensive for high-dimensional applications such as imaging. Proximal methods have low per-iteration complexity. However, its convergence speed is typically slow.

Message passing is a promising alternative to solve the BPDN problem in (2.2). To apply message passing, we first notice that (2.2) can be viewed as a maximum a posteriori probability (MAP) estimation problem. Specifically, we assign a prior

X. Yuan and Z. Xue, *Turbo Message Passing Algorithms for Structured Signal Recovery*, SpringerBriefs in Computer Science,
https://doi.org/10.1007/978-3-030-54762-2_2

distribution $p(\boldsymbol{x}) \propto \exp(\frac{-\lambda\|\boldsymbol{x}\|_1}{\sigma^2})$ to $\boldsymbol{x}$. Then, it is easy to verify that $\hat{\boldsymbol{x}}$ in (2.2) is equivalent to:

$$\hat{\boldsymbol{x}} = \arg\max_{\boldsymbol{x}\in\mathbb{R}^n} p(\boldsymbol{x}|\boldsymbol{y}) \tag{2.3}$$

where $p(\boldsymbol{x}|\boldsymbol{y}) = p(\boldsymbol{y}|\boldsymbol{x})p(\boldsymbol{x})/p(\boldsymbol{y})$.

In [3], a factor graph was established to represent above the probability model, based on which approximate message passing (AMP) was used to iteratively solve the inference problem in (2.3). As the established factor graph is dense in general, directly applying message passing to the graph leads to high complexity. To reduce complexity, two approximations are introduced in AMP: First, messages from factor nodes to variable nodes are nearly Gaussian; second, messages from variable nodes to factor nodes can be calculated by using Taylor-series approximation to reduce computational cost. It was shown in [3] that the approximation error vanishes when $m, n \to \infty$ with a fixed ratio.

The convergence of the AMP algorithm requires that the elements of the sensing matrix $\boldsymbol{A}$ are sufficiently random. It was shown in [4] that AMP is asymptotically optimal when $\boldsymbol{A}$ is independent and identically distributed (i.i.d.) Gaussian and the behavior of AMP can be characterized by a scalar recursion called state evolution.

2.1.2 Turbo Compressed Sensing

In many applications, the sensing matrix $\boldsymbol{A}$ is neither i.i.d. nor Gaussian. For example, to reduce storage and computational complexity, measurements are usually taken from an orthogonal transform domain, such as discrete Fourier transform (DFT) or discrete cosine transform (DCT). In these scenarios, the performance of AMP deteriorates and the convergence of AMP is not guaranteed. This motivates the development of the Turbo-CS algorithm [5] described below.

The block diagram of the Turbo-CS algorithm is illustrated in Fig. 2.1. Turbo-CS bears a structure similar to a turbo decoder [6], hence the name Turbo-CS. As illustrated in Fig. 2.1, the Turbo-CS algorithm consists of two modules. Module A is basically a linear minimum mean square error (LMMSE) estimator of $\boldsymbol{x}$ based on the measurement $\boldsymbol{y}$ and the messages from Module B. Module B performs minimum mean square error (MMSE) estimation that combines the prior distribution of $\boldsymbol{x}$ and the messages from Module A. The two modules are executed iteratively to refine the estimate of $\boldsymbol{x}$. The detailed operations of Turbo-CS are presented in Algorithm 1.

We now give more details of Algorithm 1. Module A estimates $\boldsymbol{x}$ based on the measurement $\boldsymbol{y}$ in (2.1) with $\boldsymbol{x}$ *a priori* distributed as $\boldsymbol{x} \sim \mathcal{N}(\boldsymbol{x}_A^{pri}, v_A^{pri}\boldsymbol{I})$. Given $\boldsymbol{y}$ with $\boldsymbol{x} \sim \mathcal{N}(\boldsymbol{x}_A^{pri}, v_A^{pri}\boldsymbol{I})$, the posterior distribution of each x_i is still Gaussian with posterior mean and variance given by [7]

Algorithm 1 Turbo-CS algorithm

Input: $\boldsymbol{A}, \boldsymbol{y}, \sigma^2, \boldsymbol{x}_A^{pri} = \boldsymbol{0}$

1: **while** the stopping criterion is not met **do**

2: $\boldsymbol{x}_A^{ext} = \boldsymbol{x}_A^{pri} + \frac{n}{m}\boldsymbol{A}^T(\boldsymbol{y} - \boldsymbol{A}\boldsymbol{x}_A^{pri})$ %Module A

3: $v_A^{ext} = \left(\frac{n}{m} - 1\right) v_A^{pri} + \frac{n}{m}\sigma^2$

4: $\boldsymbol{x}_B^{pri} = \boldsymbol{x}_A^{ext}, v_B^{pri} = v_A^{ext}$

5: $x_{B,i}^{post} = \mathrm{E}\left[x_i | x_{B,i}^{pri}\right]$ %Module B

6: $v_B^{post} = \frac{1}{n}\sum_{i=1}^{n} \mathrm{var}\left[x_i | x_{B,i}^{pri}\right]$

7: $v_A^{pri} = v_B^{ext} = \left(\frac{1}{v_B^{post}} - \frac{1}{v_B^{pri}}\right)^{-1}$

8: $\boldsymbol{x}_A^{pri} = \boldsymbol{x}_B^{ext} = v_B^{ext}\left(\frac{\boldsymbol{x}_B^{post}}{v_B^{post}} - \frac{\boldsymbol{x}_B^{pri}}{v_B^{pri}}\right)$

9: **end while**

Output: $\boldsymbol{x}_B^{post}$

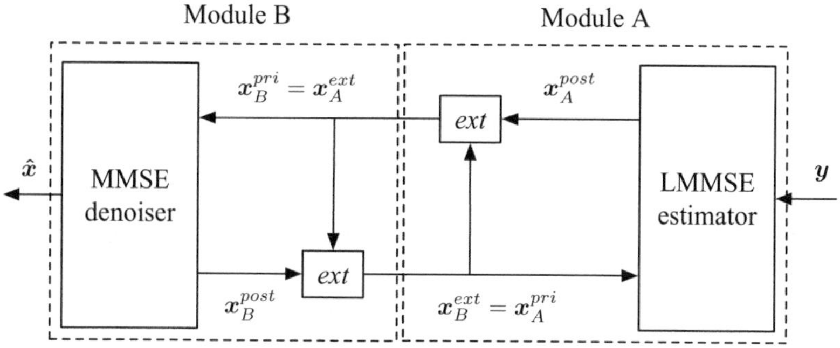

Fig. 2.1 An illustration of the Turbo-CS algorithm proposed in [5]

$$x_{A,i}^{post} = x_{A,i}^{pri} + \frac{v_A^{pri}}{v_A^{pri} + \sigma^2}\boldsymbol{a}_i^T(\boldsymbol{y} - \boldsymbol{A}\boldsymbol{x}_A^{pri}) \tag{2.4a}$$

$$v_A^{post} = v_A^{pri} - \frac{m}{n}\frac{(v_A^{pri})^2}{v_A^{pri} + \sigma^2}, \tag{2.4b}$$

where $\boldsymbol{a}_i$ is the i-th column of $\boldsymbol{A}$. Note that as the measurement $\boldsymbol{y}$ is linear in $\boldsymbol{x}$, the *a posteriori* mean in (2.4a) is also called the LMMSE estimator of x_i.

The posterior distributions cannot be used directly in message passing due to the correlation issue. Instead, we need to calculate the so-called extrinsic message [6] for each x_i by excluding the contribution of the input message of x_i. That is, the extrinsic distribution of each x_i satisfies

$$\mathcal{N}_{x_i}(x_{A,i}^{pri}, v_{A,i}^{pri})\mathcal{N}_{x_i}(x_{A,i}^{ext}, v_{A,i}^{ext}) \doteq \mathcal{N}_{x_i}(x_{A,i}^{post}, v_{A,i}^{post}), \tag{2.5}$$

where $\mathcal{N}_x(m, v) = \frac{1}{\sqrt{2\pi v}} \exp(-\frac{1}{2v}(x - m)^2)$, and "$\doteq$" represents equality up to a constant multiplicative factor. From (2.5), the extrinsic mean and variance of x_i are, respectively, given in [8] as

$$x_{A,i}^{ext} = v_A^{ext} \left(\frac{x_{A,i}^{post}}{v_A^{post}} - \frac{x_{A,i}^{pri}}{v_A^{pri}} \right) \tag{2.6a}$$

$$v_A^{ext} = \left(\frac{1}{v_A^{post}} - \frac{1}{v_A^{pri}} \right)^{-1}. \tag{2.6b}$$

Combining (2.4) and (2.6), we obtain Lines 2 and 3 of Algorithm 1.

Note that the left-hand side of (2.5), proportional to $\exp(-\frac{1}{2v_{A,i}^{pri}}(x_i - x_{A,i}^{pri})^2) \exp(-\frac{1}{2v_{A,i}^{ext}}(x_i - x_{A,i}^{ext})^2)$, gives a joint distribution of the input distortion $x_{A,i}^{pri} - x_i$ and the output distortion $x_{A,i}^{ext} - x_i$. This joint distribution implies the independence of the input and output distortion. Furthermore, for Gaussian distributions, independence is equivalent to uncorrelatedness. Thus, we have

$$\mathrm{E}\left[(x_i - x_{A,i}^{pri})(x_i - x_{A,i}^{ext})\right] = 0, \tag{2.7}$$

where the expectation is taken over the joint probability distribution of x_i, $x_{A,i}^{pri}$, and $x_{A,i}^{ext}$.

We now consider Module B. Recall that Module B estimates each x_i by combining the prior distribution $x_i \sim p(x_i)$ and the message from Module A. Note that the message $x_{A,i}^{ext}$ from Module A is now treated as an input of Module B, denoted by $x_{B,i}^{pri}$. Following [5], we model each $x_{B,i}^{pri}$ as an observation of x_i corrupted by an additive noise:

$$x_{B,i}^{pri} = x_i + n_{B,i}^{pri} \tag{2.8}$$

where $n_{B,i}^{pri} \sim \mathcal{N}(0, v_B^{pri})$ is independent of x_i. The *a posteriori* mean and variance of each x_i for Module B are, respectively, given by

$$\begin{aligned} x_{B,i}^{post} &= \mathrm{E}[x_i | x_{B,i}^{pri}] \\ v_B^{post} &= \frac{1}{n} \sum_{i=1}^{n} \mathrm{var}[x_i | x_{B,i}^{pri}], \end{aligned} \tag{2.9}$$

where $\mathrm{var}[x|y]$ denotes the conditional variance of x given y. Similar to (2.6), the extrinsic variance and mean of $\boldsymbol{x}$ for Module B are, respectively, given by Lines 7 and 8 of Algorithm 1. Also, similar to (2.7), the extrinsic distortion is uncorrelated with the prior distortion, i.e.

$$\mathrm{E}\left[(x_i - x_{B,i}^{pri})(x_i - x_{B,i}^{ext})\right] = 0, \tag{2.10}$$

where the expectation is taken over the joint probability distribution of x_i, $x_{B,i}^{pri}$, and $x_{B,i}^{ext}$. Later, we will see that (2.10) plays an important role in the extension of Turbo-CS.

2.2 Denoising-Based Turbo-CS

2.2.1 Problem Statement

In Algorithm 1, the operation of Module B requires the knowledge of the prior distribution of $\boldsymbol{x}$. However, such prior information is difficult to acquire in many applications. Low-complexity robust denoisers, rather than the optimal MMSE denoiser, are usually employed in practice, even when the prior distribution of $\boldsymbol{x}$ is available.

Turbo-CS with a generic denoiser is illustrated in Fig. 2.2. Compared with Fig. 2.1, the only difference is that Turbo-CS in Fig. 2.2 replaces the MMSE denoiser by a generic denoiser, defined as

$$\boldsymbol{x}_B^{post} = \boldsymbol{D}(\boldsymbol{x}_B^{pri}; v_B^{pri}, \boldsymbol{\theta}), \tag{2.11}$$

where $\boldsymbol{D}(\cdot)$ represents the denoising function with $\boldsymbol{x}_B^{pri}$ being the input, $\boldsymbol{x}_B^{post}$ being the output, and v_B^{pri} and $\boldsymbol{\theta}$ being the parameters. Note that the choice of $\boldsymbol{\theta}$ will be specified when a specific denoiser is involved. For brevity, we may simplify the notation $\boldsymbol{D}(\boldsymbol{x}_B^{pri}; v_B^{pri}, \boldsymbol{\theta})$ to $\boldsymbol{D}(\boldsymbol{x}_B^{pri})$ in circumstances without causing ambiguity. Also, we denote the i-th entry of $\boldsymbol{x}_B^{post}$ as

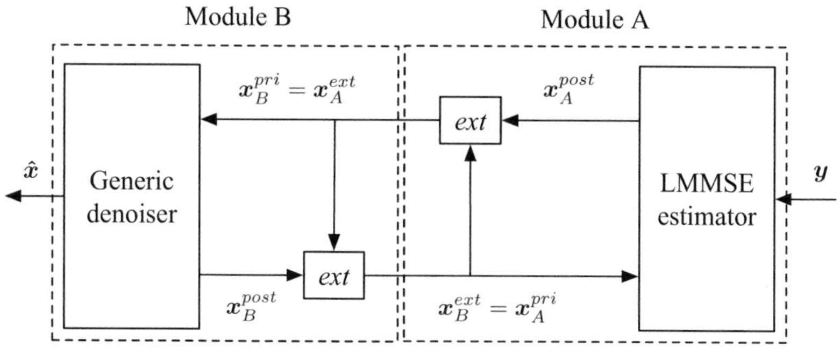

Fig. 2.2 An illustration of the denoising-based Turbo-CS algorithm

$$x_{B,i}^{post} = D_i(\boldsymbol{x}_B^{pri}). \tag{2.12}$$

With the above replacement, the main challenge is how to calculate the extrinsic message of each x_i for Module B, without the prior knowledge of the distribution $p(x_i)$. Note that Lines 7 and 8 of Algorithm 1 cannot be used any more since they hold only for the MMSE denoiser.

2.2.2 *Extrinsic Messages for a Generic Denoiser*

We now describe how to calculate the extrinsic messages for a generic denoiser. Without loss of generality, denote the extrinsic output of Module B by $\boldsymbol{x}_B^{ext} = \boldsymbol{D}^{ext}(\boldsymbol{x}_B^{pri})$. We call $\boldsymbol{D}^{ext}(\boldsymbol{x}_B^{pri})$ an extrinsic denoiser. Similarly to Line 8 of Algorithm 1, we construct $\boldsymbol{x}_B^{ext}$ by a linear combination of the *a priori* mean and the *a posteriori* mean:

$$\boldsymbol{x}_B^{ext} = \boldsymbol{D}^{ext}(\boldsymbol{x}_B^{pri}) = c(\boldsymbol{x}_B^{post} - \alpha \boldsymbol{x}_B^{pri}) \tag{2.13}$$

where c and α are coefficients to be determined. Clearly, (2.13) is identical to Line 8 of Algorithm 1 by letting $c = \frac{v_B^{ext}}{v_B^{post}}$ and $\alpha = \frac{v_B^{post}}{v_B^{pri}}$. We follow the turbo message passing principle and require that c and α are chosen such that

- The extrinsic distortion is uncorrelated with the prior distortion, i.e.

$$\mathrm{E}[(\boldsymbol{x} - \boldsymbol{x}_B^{pri})^T(\boldsymbol{x} - \boldsymbol{x}_B^{ext})] = 0; \tag{2.14}$$

- $\mathrm{E}[\|\boldsymbol{x}_B^{ext} - \boldsymbol{x}\|^2]$ is minimized.

From the discussions in Sect. 2.1.2, the calculation of the extrinsic messages in Lines 8 and 9 satisfies the above two conditions when the MMSE denoiser is employed. Note that (2.14) is a relaxation of (2.10) since (2.10) implies (2.14) but the converse does not necessarily hold. Later we will see that this relaxation is good for many applications. What remains is to determine c and α satisfying conditions 2.2.2 and 2.2.2 for a generic denoiser. This is elaborated in the following.

2.2.2.1 Determining Parameter α

As mentioned in Sect. 2.1.2, the input message of Module B can be modeled by (2.8), where the noise part $n_{B,i}^{pri}$ is independent of x_i. Then

$$\mathrm{E}[(\boldsymbol{x} - \boldsymbol{x}_B^{pri})^T(\boldsymbol{x} - \boldsymbol{x}_B^{ext})] = -\mathrm{E}[(\boldsymbol{n}_B^{pri})^T(\boldsymbol{x} - \boldsymbol{x}_B^{ext})] \tag{2.15a}$$

$$= -\sum_{i=1}^{n} \mathrm{E}\left[n_{B,i}^{pri} x_{B,i}^{ext}\right] \tag{2.15b}$$

where (2.15a) follows from (2.8), and (2.15b) follows by noting $\mathrm{E}[(\boldsymbol{n}_B^{pri})^T \boldsymbol{x}] = 0$. To proceed, we introduce the Stein's lemma [9].

Lemma 2.1 *For a normally distributed random variable* $y \sim \mathcal{N}(\mu_y, \sigma_y^2)$, *and a differentiable function* $h : \mathbb{R} \to \mathbb{R}$ *such that* $\mathrm{E}[|h'(y)|] < \infty$, *we have*

$$\sigma_y^2 \mathrm{E}[h'(y)] = \mathrm{E}[(y - \mu_y) h(y)]. \tag{2.16}$$

By applying Stein's lemma, we have

$$\begin{aligned}
\mathrm{E}\left[n_{B,i}^{pri} x_{B,i}^{ext}\right] &= c\mathrm{E}\left[n_{B,i}^{pri}(x_{B,i}^{post} - \alpha(x_i + n_{B,i}^{pri}))\right] && (2.17a)\\
&= c\mathrm{E}[n_{B,i}^{pri} x_{B,i}^{post}] - c\alpha\mathrm{E}[n_{B,i}^{pri} x_i] - c\alpha\mathrm{E}[n_{B,i}^{pri} n_{B,i}^{pri}] &&\\
&= c\mathrm{E}\left[n_{B,i}^{pri} D_i(\boldsymbol{x}_B^{pri})\right] - c\alpha v_B^{pri} && (2.17b)\\
&= c\mathrm{E}\left[(x_{B,i}^{pri} - x_i) D_i(\boldsymbol{x}_B^{pri})\right] - c\alpha v_B^{pri} && (2.17c)\\
&= c v_B^{pri} \mathrm{E}\left[D_i'(\boldsymbol{x}_B^{pri})\right] - c\alpha v_B^{pri} && (2.17d)
\end{aligned}$$

where $D_i'(\boldsymbol{x}_B^{pri})$ denotes the partial derivative of $D_i(\boldsymbol{x}_B^{pri})$ with respect to variable $x_{B,i}^{pri}$ and the expectation is taken over the probability distribution of $x_{B,i}^{pri}$. In the above, (2.17a) follows from (2.8) and (2.13), (2.17c) from $\mathrm{E}[x_{B,i}^{pri} x_i] = 0$ and $\mathrm{E}[n_{B,i}^{pri} n_{B,i}^{pri}] = v_B^{pri}$, and (2.17e) from the Stein's lemma by letting $y = x_{B,i}^{pri}$. Combining (2.14), (2.15b), and (2.17), we obtain

$$\begin{aligned}
\alpha &= \frac{1}{n}\mathrm{E}\left[\sum_{i=1}^{n} D_i'(\boldsymbol{x}_B^{pri})\right] && (2.18a)\\
&\approx \frac{1}{n}\sum_{i=1}^{n} D_i'(\boldsymbol{x}_B^{pri}) && (2.18b)\\
&= \frac{1}{n}\mathrm{div}\{\boldsymbol{D}(\boldsymbol{x}_B^{pri})\}, && (2.18c)
\end{aligned}$$

where $\mathrm{div}\{\boldsymbol{D}(\boldsymbol{x}_B^{pri})\}$ denotes divergence of denoiser $\boldsymbol{D}(\boldsymbol{x}_B^{pri})$. Note that the approximation in (2.18b) becomes accurate when n is large. Also, with this approximation, the calculation of α does not depend on the distribution of $\boldsymbol{x}$.

By substituting (2.18) into (2.13), we obtain

$$\boldsymbol{x}_B^{ext} = c\left(\boldsymbol{x}_B^{post} - \frac{1}{n}\mathrm{div}\{\boldsymbol{D}(\boldsymbol{x}_B^{pri})\}\boldsymbol{x}_B^{pri}\right) \tag{2.19a}$$

$$= c\left(\boldsymbol{D}(\boldsymbol{x}_B^{pri}) - \frac{1}{n}\mathrm{div}\{\boldsymbol{D}(\boldsymbol{x}_B^{pri})\}\boldsymbol{x}_B^{pri}\right) \tag{2.19b}$$

$$= \boldsymbol{D}^{ext}(\boldsymbol{x}_B^{pri}), \tag{2.19c}$$

where the extrinsic denoiser $\boldsymbol{D}^{ext}(\cdot)$ is defined as

$$\boldsymbol{D}^{ext}(\boldsymbol{r}) = c\left(\boldsymbol{D}(\boldsymbol{r}) - \frac{1}{n}\mathrm{div}\{\boldsymbol{D}(\boldsymbol{x}_B^{pri})\}\boldsymbol{r}\right). \tag{2.20}$$

The divergence of $\boldsymbol{D}^{ext}(\boldsymbol{r})$ at $\boldsymbol{r} = \boldsymbol{x}_B^{ext}$ is zero by noting

$$\mathrm{div}\{\boldsymbol{D}^{ext}(\boldsymbol{x}_B^{pri})\} = c\left(\mathrm{div}\{\boldsymbol{D}(\boldsymbol{x}_B^{pri})\} - \mathrm{div}\{\boldsymbol{D}(\boldsymbol{x}_B^{pri})\}\right) = 0. \tag{2.21}$$

Thus, $\boldsymbol{D}^{ext}(\cdot)$ belongs to the family of divergence-free denoisers proposed in [10].

2.2.2.2 Determining Parameter c

Ideally, we want to choose parameter c to satisfy condition 2.2.2 below (2.14). However, the MSE is difficult to evaluate as the distribution of $\boldsymbol{x}$ is unknown. To address this problem, we use the Stein's unbiased risk estimate (SURE) given in [11] to approximate the MSE.

To be specific, consider the signal model

$$\boldsymbol{r} = \boldsymbol{x} + \tau\boldsymbol{n}, \tag{2.22}$$

where $\boldsymbol{n} \in \mathbb{R}^{n\times 1}$ is the additive Gaussian noise draw from $\mathcal{N}(0, \boldsymbol{I})$. The mean square error of denoiser $\boldsymbol{D}(\boldsymbol{r})$ is defined by

$$\mathrm{MSE} = \frac{1}{n}\mathrm{E}\left[\|\boldsymbol{D}(\boldsymbol{r}) - \boldsymbol{x}\|^2\right]. \tag{2.23}$$

The SURE [11, Theorem 1] of the MSE of $\boldsymbol{D}(\boldsymbol{r})$ is given by

$$\widehat{\mathrm{MSE}} = \frac{1}{n}\|\boldsymbol{D}(\boldsymbol{r}) - \boldsymbol{r}\|^2 + \frac{2\tau^2}{n}\mathrm{div}\{\boldsymbol{D}(\boldsymbol{r})\} - \tau^2. \tag{2.24}$$

Compared with the MSE in (2.23), the SURE in (2.24) does not involve the distribution of $\boldsymbol{x}$. We next use SURE as a surrogate for MSE and tune the denoiser by minimizing the SURE. Recall from (2.8) that $\boldsymbol{x}_B^{pri}$ can be represented as $\boldsymbol{x}_B^{pri} = \boldsymbol{x} + \boldsymbol{n}_B^{pri}$. Let $\tau = \sqrt{v_B^{pri}}$. Then, applying (2.24) to $\boldsymbol{D}^{ext}(\boldsymbol{x}_B^{pri})$, we obtain

$$\begin{aligned}\widehat{\text{MSE}} =& \frac{1}{n}\|\boldsymbol{D}^{ext}(\boldsymbol{x}_B^{pri}) - \boldsymbol{x}_B^{pri}\|^2 + \frac{2v_B^{pri}}{n}\text{div}\{\boldsymbol{D}(\boldsymbol{x}_B^{pri})\} - v_B^{pri} \\ =& \frac{1}{n}\|\boldsymbol{D}^{ext}(\boldsymbol{x}_B^{pri}) - \boldsymbol{x}_B^{pri}\|^2 - v_B^{pri} \\ =& \frac{1}{n}\left\| c\left(\boldsymbol{D}(\boldsymbol{x}_B^{pri}) - \frac{1}{n}\text{div}\{\boldsymbol{D}(\boldsymbol{x}_B^{pri})\}\boldsymbol{x}_B^{pri}\right) - \boldsymbol{x}_B^{pri}\right\|^2 - v_B^{pri}\end{aligned} \tag{2.25}$$

where the second step follows from (2.21), and the last step from (2.19). Minimizing the SURE given in (2.25), we obtain the optimal c given by

$$c^{opt} = \frac{(\boldsymbol{x}_B^{pri})^T\left(\boldsymbol{D}(\boldsymbol{x}_B^{pri}) - \frac{1}{n}\text{div}\{\boldsymbol{D}(\boldsymbol{x}_B^{pri})\}\boldsymbol{x}_B^{pri}\right)}{\|\boldsymbol{D}(\boldsymbol{x}_B^{pri}) - \frac{1}{n}\text{div}\{\boldsymbol{D}(\boldsymbol{x}_B^{pri})\}\boldsymbol{x}_B^{pri}\|^2}. \tag{2.26}$$

2.2.3 Denoising-Based Turbo-CS

We are now ready to extend Turbo-CS for a generic denoiser. We refer to the extended algorithm as Denoising-based Turbo-CS (D-Turbo-CS). The details of D-Turbo-CS are presented in Algorithm 2.

Algorithm 2 D-Turbo-CS algorithm

Input: $\boldsymbol{A}, \boldsymbol{y}, \sigma^2, \boldsymbol{x}_A^{pri} = 0$
1: **while** the stopping criterion is not met **do**
2: $\quad \boldsymbol{x}_A^{ext} = \boldsymbol{x}_A^{pri} + \frac{n}{m}\boldsymbol{A}^T(\boldsymbol{y} - \boldsymbol{A}\boldsymbol{x}_A^{pri})$ %Module A
3: $\quad v_A^{ext} = \left(\frac{n}{m} - 1\right)v_A^{pri} + \frac{n}{m}\sigma^2$
4: $\quad \boldsymbol{x}_B^{pri} = \boldsymbol{x}_A^{ext}, v_B^{pri} = v_A^{ext}$
5: $\quad \boldsymbol{x}_B^{post} = \boldsymbol{D}(\boldsymbol{x}_B^{pri}; v_B^{pri}, \boldsymbol{\theta})$ %Module B
6: $\quad \boldsymbol{x}_B^{ext} = c^{opt}(\boldsymbol{x}_B^{post} + \alpha\boldsymbol{x}_B^{pri})$
7: $\quad v_B^{ext} = \frac{\|\boldsymbol{y} - \boldsymbol{A}\boldsymbol{x}_B^{ext}\|^2 - m\sigma^2}{m}$
8: $\quad \boldsymbol{x}_A^{pri} = \boldsymbol{x}_B^{ext}, v_A^{pri} = v_B^{ext}$
9: **end while**
Output: $\boldsymbol{x}_B^{post}$

Compared with Turbo-CS, D-Turbo-CS has the same operations in Module A. But for Module B, D-Turbo-CS employs a generic denoiser, rather than the MMSE denoiser. Correspondingly, the extrinsic mean is calculated using Line 6 of Algorithm 2; the extrinsic variance is calculated in Line 7 by following Eqn. (71) in [12].

2.3 Construction of Extrinsic Denoisers

Various denoisers have been proposed in the literature for noise suppression. For example, the SURE-LET [11], the BM3D [13], and the dictionary learning [14] are developed for image denoising. In this section, we study the applications of these denoisers in D-Turbo-CS. We describe how to construct the corresponding extrinsic denoiser $\boldsymbol{D}^{ext}(\boldsymbol{r};\boldsymbol{\theta})$ for any given denoiser $\boldsymbol{D}(\boldsymbol{r};\boldsymbol{\theta})$. Based on that, we further consider optimizing the denoiser parameter $\boldsymbol{\theta}$.

2.3.1 *Extrinsic SURE-LET Denoiser*

We start with the SURE-LET denoiser. A SURE-LET denoiser is constructed as a linear combination of some kernel functions. The combination coefficients are determined by minimizing the SURE of the MSE [11].

Specifically, a SURE-LET denoiser is constructed as

$$\boldsymbol{D}(\boldsymbol{r};\boldsymbol{\theta}) = \sum_{k=1}^{K} \theta_k \boldsymbol{O}\boldsymbol{\psi}_k(\boldsymbol{O}^T\boldsymbol{r}) \tag{2.27a}$$

$$= \sum_{i=1}^{K} \theta_k \boldsymbol{\Psi}_k(\boldsymbol{r}), \tag{2.27b}$$

where $\boldsymbol{O} \in \mathbb{R}^{n\times n}$ is an orthonormal transform matrix, $\boldsymbol{\psi}_k : \mathbb{R}^n \to \mathbb{R}^n$ for $k = 1, \cdots, K$ are kernel functions, $\boldsymbol{\theta} = [\theta_1, \theta_2, \cdots, \theta_K]^T$, and $\boldsymbol{\Psi}_k(\boldsymbol{r}) = \boldsymbol{O}\boldsymbol{\psi}_k(\boldsymbol{O}^T\boldsymbol{r})$. $\boldsymbol{O}$ can be the Haar wavelet transform matrix or the DCT transform matrix.

The choice of kernel functions $\{\boldsymbol{\psi}_k\}$ depends on the structure of the input signals. For example, the authors in [15] proposed the following piecewise linear kernel functions for sparse signals:

$$\psi_{1,i}(\boldsymbol{r}) = \begin{cases} 0 & r_i \le -2\beta_1, r_i \ge 2\beta_1 \\ -\frac{r_i}{\beta_1} - 2 & -2\beta_1 < r_i < -\beta_1 \\ \frac{r_i}{\beta_1} & -\beta_1 \le r_i \le \beta_1 \\ -\frac{r_i}{\beta_1} + 2 & \beta_1 < r_i < 2\beta_1 \end{cases} \tag{2.28}$$

$$\psi_{2,i}(\boldsymbol{r}) = \begin{cases} -1 & r_i \le -\beta_2 \\ \frac{r_i+\beta_1}{\beta_2-\beta_1} & -\beta_2 < r_i < -\beta_1 \\ 0 & -\beta_1 \le r_i \le \beta_1 \\ \frac{r_i-\beta_1}{\beta_2-\beta_1} & \beta_1 < r_i < \beta_2 \\ 1 & r_i \ge \beta_2 \end{cases} \tag{2.29}$$

$$\psi_{3,i}(\boldsymbol{r}) = \begin{cases} r_i + \beta_2 & r_i \le -\beta_2 \\ 0 & -\beta_2 < r_i < \beta_2 \\ r_i - \beta_2 & r_i \ge \beta_2 \end{cases} \tag{2.30}$$

where $\psi_{k,i}(\boldsymbol{r})$ represents the i-th element of $\boldsymbol{\psi}_k(\boldsymbol{r})$, r_i is the i-th element of $\boldsymbol{r}$, β_1 and β_2 are constants chosen based on the noise level τ^2. The recommended values of β_1 and β_2 can be found in [15].

For SURE-LET denoiser $\boldsymbol{D}(\boldsymbol{r};\boldsymbol{\theta})$ in (2.27), the corresponding extrinsic denoiser $\boldsymbol{D}^{ext}(\boldsymbol{r};\boldsymbol{\theta})$ is given by

$$\boldsymbol{D}^{ext}(\boldsymbol{r};\boldsymbol{\theta}) = c\left(\sum_{i=1}^{K}\theta_k\boldsymbol{\Psi}_k(\boldsymbol{r}) - \frac{1}{n}\text{div}\left\{\sum_{i=1}^{K}\theta_k\boldsymbol{\Psi}_k(\boldsymbol{r})\right\}\boldsymbol{r}\right) \tag{2.31a}$$

$$= \sum_{i=1}^{K}\theta'_k\left(\boldsymbol{\Psi}_k(\boldsymbol{r}) - \frac{1}{n}\text{div}\{\boldsymbol{\Psi}_k(\boldsymbol{r})\}\boldsymbol{r}\right), \tag{2.31b}$$

where (2.31a) is from (2.20), and $\theta'_k = c\theta_k$, for $k = 1, \cdots, K$.

We next determine the optimal $\boldsymbol{\theta}' = [\theta'_1, \cdots, \theta'_K]^T$ by minimizing the SURE. From (2.24), the SURE of $\boldsymbol{D}^{ext}(\boldsymbol{r},\boldsymbol{\theta})$ is given by

$$\widehat{\text{MSE}} = \frac{1}{n}\|\boldsymbol{D}^{ext}(\boldsymbol{r},\boldsymbol{\theta}) - \boldsymbol{r}\|^2 + \frac{2\tau^2}{n}\text{div}\{\boldsymbol{D}^{ext}(\boldsymbol{r})\} - \tau^2 \tag{2.32a}$$

$$= \frac{1}{n}\left\|\sum_{i=1}^{K}\theta'_k\left(\boldsymbol{\Psi}_k(\boldsymbol{r}) - \frac{1}{n}\text{div}\{\boldsymbol{\Psi}_k(\boldsymbol{r})\}\boldsymbol{r}\right) - \boldsymbol{r}\right\|^2 - \tau^2 \tag{2.32b}$$

$$= \frac{1}{n}\left\|\sum_{k=1}^{K}\theta'_k\left(\boldsymbol{\psi}_k(\tilde{\boldsymbol{r}}) - \frac{1}{n}\text{div}\{\boldsymbol{\psi}_k(\tilde{\boldsymbol{r}})\}\tilde{\boldsymbol{r}}\right) - \tilde{\boldsymbol{r}}\right\|^2 - \tau^2 \tag{2.32c}$$

where (2.32b) follows from (2.21) and (2.31), (2.32c) follows from $\boldsymbol{\Psi}_k(\boldsymbol{O}\tilde{\boldsymbol{r}}) = \boldsymbol{O}\boldsymbol{\psi}_k(\tilde{\boldsymbol{r}})$ and $\text{div}\{\boldsymbol{\Psi}_k(\boldsymbol{r})\} = \text{div}\{\boldsymbol{\psi}_k(\tilde{\boldsymbol{r}})\}$ with $\tilde{\boldsymbol{r}} = \boldsymbol{O}^T\boldsymbol{r}$.

The optimal $\boldsymbol{\theta}'$ that minimizes $\widehat{\text{MSE}}$ in (2.32) is given by

$$(\boldsymbol{\theta}')^{opt} = \boldsymbol{M}^{-1}\boldsymbol{b}, \tag{2.33}$$

where the (i, j)th entry of $\boldsymbol{M} \in \mathbb{R}^{K\times K}$ and the i-th entry of $\boldsymbol{b} \in \mathbb{R}^{K\times 1}$ are, respectively, given by

$$M_{i,j} = [\boldsymbol{\psi}_i^{ext}(\tilde{\boldsymbol{r}})]^T\boldsymbol{\psi}_j^{ext}(\tilde{\boldsymbol{r}}) \tag{2.34a}$$

$$b_i = [\boldsymbol{\psi}_i^{ext}(\tilde{\boldsymbol{r}})]^T\tilde{\boldsymbol{r}}, \tag{2.34b}$$

with

$$\boldsymbol{\psi}_k^{ext}(\tilde{\boldsymbol{r}}) = \boldsymbol{\psi}_k(\tilde{\boldsymbol{r}}) - \frac{1}{n}\text{div}\{\boldsymbol{\psi}_k(\tilde{\boldsymbol{r}})\}\tilde{\boldsymbol{r}}. \tag{2.35}$$

2.3.2 Other Extrinsic Denoisers

The SURE-LET denoiser has analytical expressions. However, there are other denoisers that cannot be expressed in a closed form. The corresponding extrinsic denoisers also have no analytical expressions. We give two examples as follows.

The first example is the dictionary learning denoiser. Dictionary learning aims to find a sparse representation for a given data set in the form of a linear combination of a set of basic elements. This set of basic elements is called a dictionary. Existing dictionary learning algorithms include K-SVD [16], iterative least square (ILS) [17], recursive least squares (RLS) [18], and the sequential generalization of K-means (SGK) [19]. Based on the above dictionary learning algorithms, we can construct dictionary learning denoisers by following the approach in [14]. Specifically, consider a noisy image matrix $\boldsymbol{R} \in \mathbb{R}^{n_1 \times n_2}$, where n_1 and n_2 are integers. We reshape $\boldsymbol{R}$ into a vector $\boldsymbol{r} \in \mathbb{R}^{n\times 1}$, where $n = n_1 n_2$. Also, we divide the whole image into blocks of size $n_3 \times n_3$, and reshape each block $\boldsymbol{R}_{i,j}$ into a vector $\boldsymbol{r}_{i,j}$, where n_3 is an integer satisfying $n_3 \ll n_1, n_2$. Note that $\boldsymbol{r}_{i,j}$ is related to $\boldsymbol{r}$ by $\boldsymbol{r}_{i,j} = \boldsymbol{E}_{i,j}\boldsymbol{r}$ where $\boldsymbol{E}_{i,j} \in R^{n_3^2 \times n}$ is the corresponding block extraction matrix. Then we use $\{\boldsymbol{r}_{i,j}\}$ as the training set to train a dictionary $\boldsymbol{Q} \in n_3^2 \times n_4$ using any of the dictionary learning algorithms mentioned above, where n_4 is an integer satisfying $n_4 > n_3^2$. The image block $\boldsymbol{r}_{i,j}$ can be expressed approximately as

$$\boldsymbol{r}_{i,j} = \boldsymbol{Q}\boldsymbol{\alpha}_{i,j}, \tag{2.36}$$

where $\boldsymbol{\alpha}_{i,j} \in \mathbb{R}^{n_4\times 1}$ is the sparse representation of $\boldsymbol{r}_{i,j}$ using the dictionary $\boldsymbol{Q}$. Then, we update the whole image vector $\boldsymbol{r}$ based on the learned dictionary $\boldsymbol{Q}$ and coefficients $\boldsymbol{\alpha}_{i,j}$ by averaging the denoised image block vectors as

$$\tilde{\boldsymbol{r}} = \left(\lambda \boldsymbol{I} + \sum_{i,j} \boldsymbol{E}_{i,j}^T \boldsymbol{E}_{i,j}\right)^{-1} \left(\lambda \boldsymbol{r} + \sum_{i,j} \boldsymbol{E}_{i,j}^T \boldsymbol{Q}\boldsymbol{\alpha}_{i,j}\right), \tag{2.37}$$

where λ is a constant depending on the input noise level. Finally, we reshape the image vector $\tilde{\boldsymbol{r}}$ back into an image matrix.

The second example is the BM3D denoiser [13]. The denoising process of BM3D is summarized as follows. First, the image matrix $\boldsymbol{R}$ is separated into image blocks of size $s_1 \times s_1$ (with $7 \leq s_1 \leq 13$). For each image block, similar blocks are found and grouped together into a three-dimensional (3D) data array. Then, collaborative filtering is used to denoise the 3D data arrays. The filtered blocks are then returned back to their original positions. Note that BM3D achieves the state-of-the-art visual quality among all the existing image denoisers.

The above dictionary learning and BM3D denoisers have no closed-form expressions, and so the divergences of these denoisers cannot be calculated explicitly. Instead, we evaluate their divergences using the Monte Carlo method. Specifically, the divergence of $\boldsymbol{D}(\boldsymbol{R})$ can be estimated by

$$\operatorname{div}\{\boldsymbol{D}(\boldsymbol{R})\} \approx E_{\tilde{\boldsymbol{N}}}\left[\left\langle \tilde{\boldsymbol{N}}, \left(\frac{\boldsymbol{D}(\boldsymbol{R}+\delta\tilde{\boldsymbol{N}})-\boldsymbol{D}(\boldsymbol{R})}{\delta}\right)\right\rangle\right], \tag{2.38}$$

where δ is a small constant, $\tilde{\boldsymbol{N}} \in \mathbb{R}^{n_1 \times n_2}$ is a perturbation matrix with the elements i.i.d. drawn from $\mathcal{N}(0, 1)$, and $\langle \boldsymbol{A}, \boldsymbol{B} \rangle = \sum_{i,j} A_{i,j} B_{i,j}$ with $A_{i,j}$ and $B_{i,j}$ be the (i, j)th elements of $\boldsymbol{A}$ and $\boldsymbol{B}$, respectively. The expectation in (2.38) can be approximated by sample average. It is observed in [20] that one sample is good enough for high-dimensional problems.

2.4 Evolution Analysis of D-Turbo-CS

2.4.1 MSE Evolution

The behavior of D-Turbo-CS can be characterized by the so-called MSE evolution. Denote the input normalized mean square error (NMSE) of Module A (or equivalently, the output NMSE of Module B) at iteration t as $v(t)$, and the output NMSE of Module A (or equivalently, the input NMSE of Module B) at iteration t as $\tau^2(t)$, where NMSE is defined by

$$\text{NMSE} = \frac{\|\hat{\boldsymbol{x}} - \boldsymbol{x}\|_2^2}{\|\boldsymbol{x}\|_2^2}. \tag{2.39}$$

Then, the MSE evolution is characterized by

$$\tau^2(t) = \left(\frac{n}{m} - 1\right) v(t) + \frac{n}{m}\sigma^2 \tag{2.40a}$$

$$v(t+1) = \frac{1}{n}\mathrm{E}\left[\left\|\boldsymbol{D}^{ext}\left(\boldsymbol{x} + \tau(t)\boldsymbol{e}\right) - \boldsymbol{x}\right\|_2^2\right], \tag{2.40b}$$

where the (2.40a) follows from Line 3 of Algorithm 2, (2.40b) follows from the assumption in (2.8), the expectation in (2.40b) is taken over $\boldsymbol{e} \sim \mathcal{N}(\boldsymbol{0}, \boldsymbol{I})$, and $v(0)$ is initialized as $\mathrm{E}[\|\boldsymbol{x}\|_2^2]/n$. We next examine the accuracy of the above MSE evolution.

2.4.2 Sparse Signal with i.i.d. Entries

We consider the situation of $\boldsymbol{x}$ with i.i.d. entries. In simulation, each x_i in $\boldsymbol{x}$ is Gaussian-Bernoulli distributed with probability density function $p(x_i) = (1 - \rho)\delta(x_i) + \rho\mathcal{N}(x_i, 0, 1/\rho)$, where $\delta(\cdot)$ is the Dirac delta function. The other settings are: the sparsity rate $\rho = 0.27$, the measurement rate $m/n = 0.5$, the signal length $n = 20000$, and the sensing matrix is chosen as the random partial DCT defined by

$$\boldsymbol{A}_1 = \boldsymbol{S}\boldsymbol{W} \tag{2.41}$$

where $\boldsymbol{S} \in \mathbb{R}^{m \times n}$ is a random row selection matrix which consists of randomly selected rows from a permutation matrix, and $\boldsymbol{W} \in \mathbb{R}^{n \times n}$ is the DCT matrix. In simulation, the SURE-LET denoiser with the kernel functions given in (2.28)–(2.30) is employed in D-Turbo-CS and D-AMP, with the corresponding algorithms denoted by LET-Turbo-CS and LET-AMP, respectively.

As shown in Fig. 2.3, the MSE evolution of LET-Turbo-CS matches well with the simulation. In contrast, for LET-AMP, the state evolution deviates from the simulation. Also, LET-Turbo-CS outperforms LET-AMP[1] considerably and

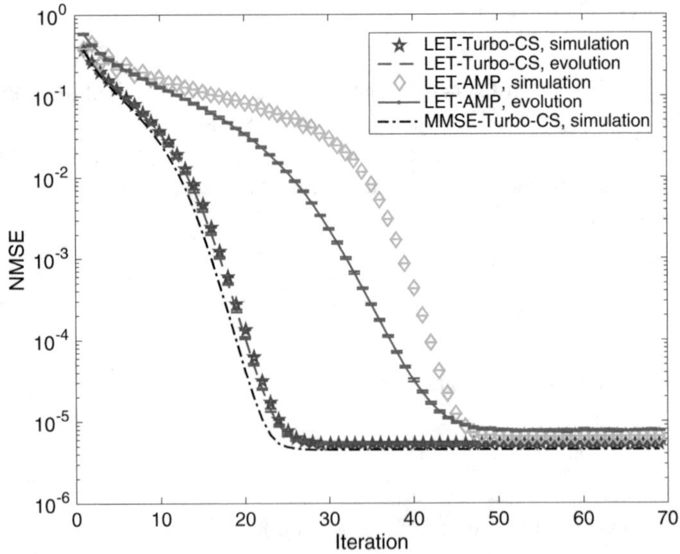

Fig. 2.3 The MSE comparison of LET-Turbo-CS and LET-AMP with the sensing matrix given by (2.41)

[1]The performance of LET-AMP here is better than the original LET-AMP [15] since here we replace the estimated variance $\hat{\sigma}^2$ in D-AMP with a more robust estimate $\hat{\sigma}^2 = \sqrt{\frac{1}{\ln 2}}\text{median}(|\hat{x}|)$ given in [21].

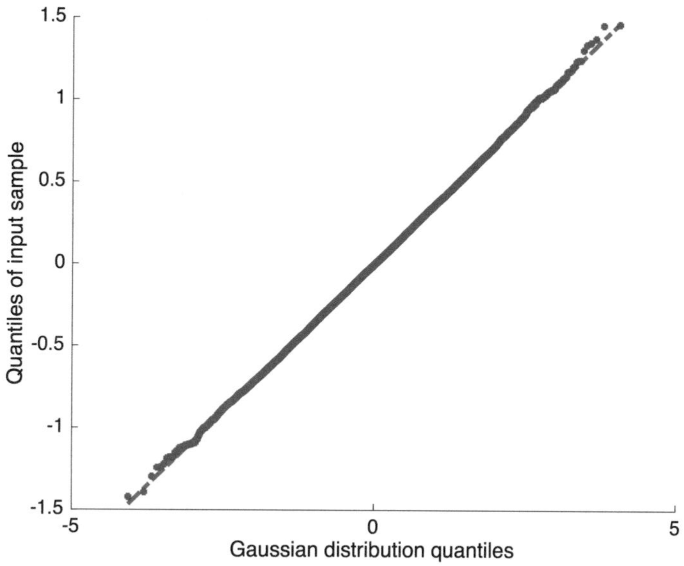

Fig. 2.4 The QQ plot of the estimation error $\boldsymbol{x}_B^{pri} - \boldsymbol{x}$ in the 10th iteration of the LET-Turbo-CS algorithm with the sensing matrix given by (2.41)

performs close to MMSE-Turbo-CS in which the MMSE denoiser is employed. We also plot the QQ plot of the estimation error of $\boldsymbol{x}_B^{pri}$ at iteration 10 of LET-Turbo-CS in Fig. 2.4. From the QQ plot, we see that $\boldsymbol{x}_B^{pri} - \boldsymbol{x}$ is close to zero mean Gaussian, which agrees well with the assumption in (2.8). Later, we will see that the Gaussianity of $\boldsymbol{x}_B^{pri} - \boldsymbol{x}$ is a good indicator of the accuracy of the MSE evolution.

2.4.3 *Signal with Correlated Entries*

In many applications, signals are correlated and the prior distribution is unknown. For example, the adjacent pixels of a natural image are correlated and their distributions are not available. We next study the MSE evolution of D-Turbo-CS for compressive image recovery.

In simulation, we generate signal $\boldsymbol{x}$ from the image "Fingerprint" of size 512×512 taken from the Javier Portilla's dataset [22] by reshaping the image into a vector of size 262144×1. The denoiser is chosen as the BM3D denoiser, and the corresponding algorithms are denoted as BM3D-Turbo-CS and BM3D-AMP. We set the measurement rate m/n to 0.3.

With the sensing matrix given in (2.41), the performance of BM3D-Turbo-CS and BM3D-AMP is simulated and shown in Fig. 2.5. We see that the simulation results of both algorithms do not match with the MSE evolution. Also, we plot the QQ plot of the estimation error $\boldsymbol{x}_B^{pri} - \boldsymbol{x}$ in Fig. 2.6. We see that the distribution of

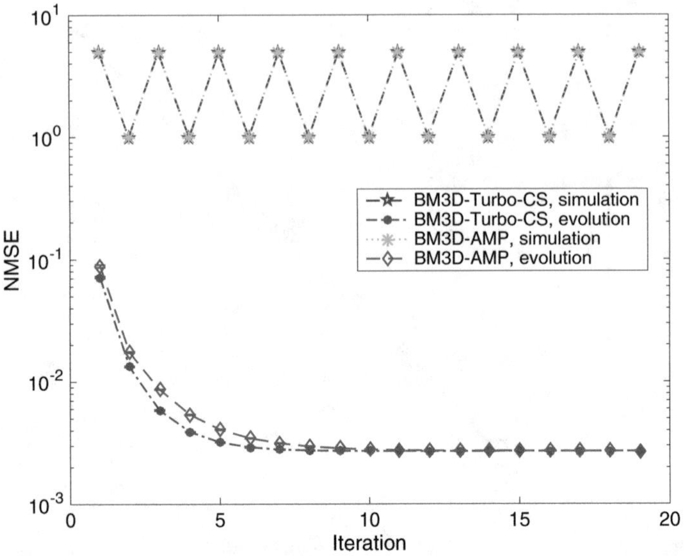

Fig. 2.5 The MSE comparison of BM3D-Turbo-CS and BM3D-AMP with the sensing matrix given by (2.41)

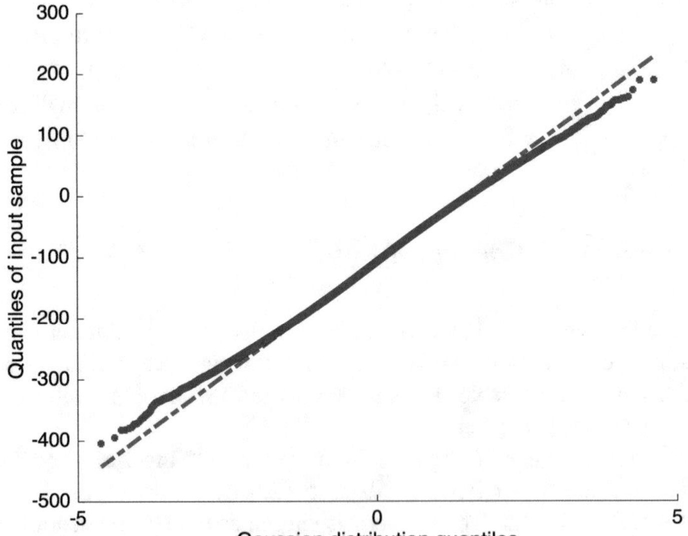

Fig. 2.6 The QQ plot of the estimation error $\boldsymbol{x}_B^{pri} - \boldsymbol{x}$ in the 2nd iteration of the BM3D-Turbo-CS algorithm with the sensing matrix given by (2.41)

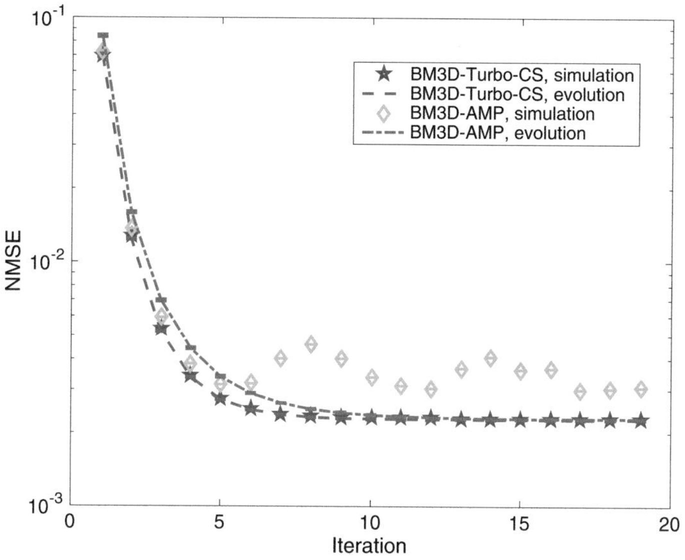

Fig. 2.7 The MSE comparison of BM3D-Turbo-CS and BM3D-AMP with the sensing matrix given by (2.42)

$\boldsymbol{x}_A^{pri} - \boldsymbol{x}$ is not quite Gaussian, and the mean of the distribution deviates from zero. This interprets the failure of the evolution prediction.

We conjecture that the reason for the degradation of the simulation performance in Fig. 2.5 is that the correlation in $\boldsymbol{x}$ is not appropriately handled. So, we replace $\boldsymbol{A}_1$ by

$$\boldsymbol{A}_2 = \boldsymbol{SW\Theta} \tag{2.42}$$

where $\boldsymbol{\Theta}$ is a diagonal matrix with the random signs (1 or -1) in the diagonal. The simulation result with sensing matrix $\boldsymbol{A}_2$ is shown in Fig. 2.7. We see that now, the MSE evolution of BM3D-Turbo-CS matches well with the simulation. Also, BM3D-Turbo-CS outperforms BM3D-AMP in both converge rate and recovery quality. In Fig 2.8, the QQ plot of the estimation error of $\boldsymbol{x}_B^{pri}$ at iteration 2 of BM3D-Turbo-CS is plotted. We see that the estimation error is close to zero mean Gaussian, similarly to the case of i.i.d. $\boldsymbol{x}$. To summarize, the sensing matrix in (2.41) is good for i.i.d. $\boldsymbol{x}$, while the sensing matrix in (2.42) is needed when the entries of $\boldsymbol{x}$ are correlated.

2.5 Performance Comparisons

In this section, we provide numerical results of D-Turbo-CS for compressive image recovery. For comparison, the recovery accuracy is measured by peak signal-to-

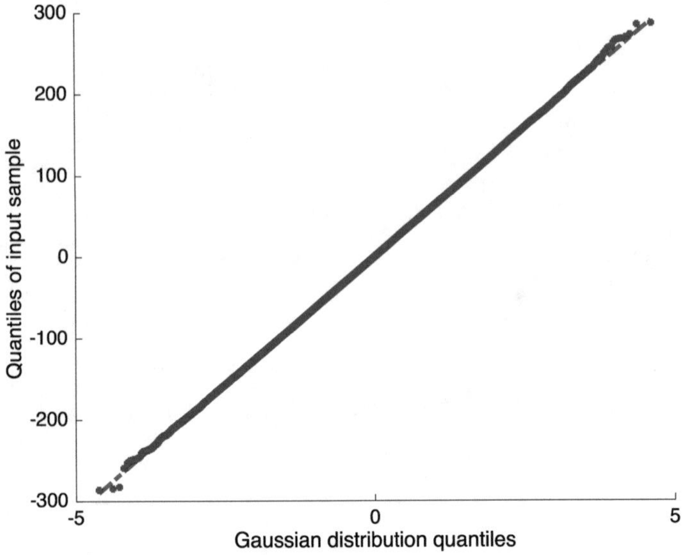

Fig. 2.8 The QQ plot of the estimation error $\boldsymbol{x}_B^{pri} - \boldsymbol{x}$ in the 2nd iteration of the LET-Turbo-CS algorithm with the sensing matrix given by (2.42)

noise ratio (PSNR):

$$\text{PNSR} = 10\log_{10}\left(\frac{\text{MAX}^2}{\text{MSE}}\right), \tag{2.43}$$

where MAX denotes the maximum possible pixel value of the image.

The stopping criterion of D-Turbo-CS is described as follows. The D-Turbo-CS algorithm stops when its output at iteration t $\hat{\boldsymbol{x}}(t)$ satisfies $\frac{\|\hat{\boldsymbol{x}}(t)-\hat{\boldsymbol{x}}(t-1)\|^2}{\|\hat{\boldsymbol{x}}(t-1)\|^2} \leq \epsilon$ or when it is executed for over T iterations, where ϵ and T are predetermined constants.

For noiseless compressive image recovery, we consider three denoisers mentioned in Sect. 2.3: the SURE-LET denoiser, the BM3D denoiser, and the dictionary learning denoiser. The corresponding algorithms of D-Turbo-CS and D-AMP are denoted by LET-Turbo-CS and LET-AMP, BM3D-Turbo-CS and BM3D-AMP, and SGK-Turbo-CS and SGK-AMP. The EM-GM-AMP algorithm in [12] is also included for comparison. The test images are chosen from the Javier Portilla's dataset, including "Lena," "Boat," "Barbara," and "Fingerprint" in Fig. 2.9. The settings of ϵ and T are as follows: $\epsilon = 10^{-4}$ and $T = 20$ for SURE-LET; $\epsilon = 10^{-4}$ and $T = 30$ for BM3D; $\epsilon = 10^{-4}$ and $T = 20$ for SGK.

In Table 2.1, we compare D-Turbo-CS with D-AMP and EM-GM-AMP for noiseless natural image recovery with the sensing matrix given in (2.42). We see that D-Turbo-CS outperforms D-AMP and EM-GM-AMP for all the test images under almost all measurement rates and denoisers. To compare the reconstruction

Table 2.1 The PSNR of the reconstructed images with sensing matrix $\boldsymbol{A}_2$

Image name	Lena						Boat					
Measurement rate	5%	10%	20%	30%	50%	70%	5%	10%	20%	30%	50%	70%
EM-GM-AMP [12]	21.56	23.33	25.22	26.89	29.50	32.38	19.55	21.06	22.87	24.70	27.78	30.95
LET-AMP [15]	–	–	–	22.09	31.38	34.57	–	–	0.23	20.02	29.62	33.28
LET-Turbo-CS	**22.27**	**24.32**	**26.77**	**28.57**	**31.74**	**35.48**	**19.96**	**21.86**	**24.43**	**26.51**	**30.16**	**34.22**
BM3D-AMP [20]	28.44	31.77	33.90	34.36	38.55	39.48	26.15	28.83	31.67	33.54	35.46	39.21
BM3D-Turbo-CS	**29.28**	**31.88**	**34.35**	**35.88**	**38.60**	**42.64**	**26.15**	**29.03**	**32.32**	**34.30**	**37.32**	**40.92**
SGK-AMP [23]	7.67	8.27	27.92	29.85	33.17	35.87	5.35	5.53	25.49	28.07	31.42	34.57
SGK-Turbo-CS	**7.70**	**8.35**	**29.01**	**31.30**	**34.60**	**37.85**	**5.39**	**5.56**	**26.22**	**28.85**	**32.54**	**35.90**
Image name	Barbara						Fingerprint					
EM-GM-AMP [12]	18.54	20.56	22.65	24.47	27.69	32.14	16.81	18.24	20.37	22.51	26.04	29.58
LET-AMP [15]	–	–	–	19.92	27.57	31.15	–	–	–	17.78	29.05	33.46
LET-Turbo-CS	**18.87**	**20.65**	**22.86**	**24.58**	**28.07**	**32.36**	**16.03**	**18.03**	**22.09**	**24.05**	**29.83**	**34.90**
BM3D-AMP [20]	**26.74**	29.52	32.81	35.21	38.46	41.66	18.18	22.75	26.61	28.59	32.07	36.44
BM3D-Turbo-CS	26.73	**30.40**	**34.23**	**36.46**	**39.91**	**43.37**	**18.04**	**24.53**	**27.73**	**30.28**	**34.51**	**38.91**
SGK-AMP [23]	**5.94**	6.35	25.58	28.20	32.13	35.78	**4.59**	4.76	20.32	23.98	**28.49**	32.39
SGK-Turbo-CS	5.88	**6.36**	**26.43**	**29.30**	**33.65**	**37.77**	4.58	**4.82**	**20.46**	**24.10**	28.38	**32.79**

The highest values of PSNR with the same denoiser are highlighted with a bold font

Table 2.2 The recovery time of different images for sensing matrix A_2. The unit of time is second

Image name	Lena						Boat					
Measurement rate	5%	10%	20%	30%	50%	70%	5%	10%	20%	30%	50%	70%
LET-AMP [15]	3.14	2.43	2.57	2.41	2.46	2.58	2.91	2.35	2.42	2.47	3.65	3.37
LET-Turbo-CS	**1.40**	**1.23**	**1.23**	**1.19**	**0.93**	**0.85**	**0.97**	**1.16**	**1.21**	**1.39**	**1.63**	**1.27**
Image name	Barbara						Fingerprint					
LET-AMP [15]	3.48	3.55	3.11	3.86	3.58	3.24	4.41	3.61	3.41	3.30	3.10	4.08
LET-Turbo-CS	**1.44**	**1.58**	**1.37**	**1.41**	**1.75**	**1.46**	**1.30**	**2.40**	**3.08**	**3.02**	**1.69**	**1.46**

The least values of time are highlighted with a bold font

Table 2.3 The PSNR of the reconstructed images for different sensing matrices

Sensing matrices	A_1			A_2			A_1			A_2		
Image name	Lena						Boat					
Measurement rate	30%	50%	70%	30%	50%	70%	30%	50%	70%	30%	50%	70%
BM3D-AMP [20]	**7.68**	7.59	13.83	34.36	38.55	39.48	–	9.23	19.11	33.54	35.46	39.21
BM3D-Turbo-CS	**7.68**	**8.69**	**13.84**	**35.88**	**38.60**	**42.64**	–	**9.32**	**19.15**	**34.30**	**37.32**	**40.92**
Image name	Barbara						Fingerprint					
BM3D-AMP [20]	**5.88**	9.93	21.06	35.21	38.46	41.66	–	7.9	20.05	28.59	32.07	36.44
BM3D-Turbo-CS	**5.88**	**10.18**	**21.07**	**36.46**	**39.91**	**43.37**	–	**8.39**	**20.07**	**30.28**	**34.51**	**38.91**

The highest values of PSNR are highlighted with a bold font

speed, we further report the reconstruction time of LET-AMP and LET-Turbo-CS in Table 2.2. Both algorithms are run until the stopping criterion is activated. We see that the reconstruction time of LET-Turbo-CS is much less than that of LET-AMP. In Table 2.3, we list the PSNR of reconstructed images using BM3D-AMP and BM3D-Turbo-CS for sensing matrix A_1 and A_2. From the table, we see that the recovery quality for sensing matrix A_1 is very poor, which is consistent with the observation in Fig. 2.5. To summarize, D-Turbo-CS has significant advantages over D-AMP and EM-GM-AMP in compressive image recovery in both visual quality and recovery time.

2.6 Summary

In this chapter, we developed the D-Turbo-CS algorithm for compressed sensing. We discussed how to construct and optimize the so-called extrinsic denoisers for D-Turbo-CS. D-Turbo-CS does not require prior knowledge of the signal distribution, and so can be adopted in many applications including compressive image recovery. Numerical results show that D-Turbo-CS outperforms D-AMP and EM-GM-AMP in terms of both recovery accuracy and convergence speed when partial orthogonal sensing matrices are involved.

Fig. 2.9 The test images of size 512×512

References

1. A.S. Nemirovski, M.J. Todd, Interior-point methods for optimization. Acta Numerica **17**, 191–234 (2008)
2. N. Parikh, S.P. Boyd et al., Proximal algorithms. Found. Trends Optim. **1**(3), 127–239 (2014)
3. D.L. Donoho, A. Maleki, A. Montanari, Message passing algorithms for compressed sensing: I. motivation and construction, in *Proc. of IEEE Information Theory Workshop (ITW)*, pp. 1–5, Cairo, Egypt, Jan. 2010
4. M. Bayati, A. Montanari, The dynamics of message passing on dense graphs, with applications to compressed sensing. IEEE Trans. Inf. Theory **57**(2), 764–785 (2011)
5. J. Ma, X. Yuan, L. Ping, Turbo compressed sensing with partial DFT sensing matrix. IEEE Signal Process. Lett. **22**(2), 158–161 (2015)
6. C. Berrou, A. Glavieux, Near optimum error correcting coding and decoding: Turbo-codes. IEEE Trans. Commun. **44**(10), 1261–1271 (1996)

7. S.M. Kay, *Fundamentals of Statistical Signal Processing: Practical Algorithm Development*, vol. 3 (Pearson Education, 2013)
8. Q. Guo, D.D. Huang, A concise representation for the soft-in soft-out LMMSE detector. IEEE Commun. Lett. **15**(5), 566–568 (2011)
9. C.M. Stein, Estimation of the mean of a multivariate normal distribution. Ann. Stat. **9**(6), 1135–1151 (1981)
10. J. Ma, L. Ping, Orthogonal AMP. IEEE Access **5**(14), 2020–2033 (2017)
11. T. Blu, F. Luisier, The SURE-LET approach to image denoising. IEEE Trans. Image Process. **16**(11), 2778–2786 (2007)
12. J.P. Vila, P. Schniter, Expectation-maximization Gaussian-mixture approximate message passing. IEEE Trans. Signal Process. **61**(19), 4658–4672 (2013)
13. K. Dabov, A. Foi, V. Katkovnik, K. Egiazarian, Image denoising with block-matching and 3D filtering, in *Proc. of Electronic Imaging (EI)*, San Jose, CA, USA, Jan. 2006
14. M. Elad, M. Aharon, Image denoising via sparse and redundant representations over learned dictionaries. IEEE Trans. Image Process. **15**(12), 3736–3745 (2006)
15. C. Guo, M.E. Davies, Near optimal compressed sensing without priors: Parametric SURE approximate message passing. IEEE Trans. Signal Process. **63**(8), 2130–2141 (2015)
16. M. Aharon, M. Elad, A. Bruckstein, K-SVD: An algorithm for designing overcomplete dictionaries for sparse representation. IEEE Trans. Signal Process. **54**(11), 4311–4322 (2006)
17. K. Engan, K. Skretting, J.H. Husoy, Family of iterative LS-based dictionary learning algorithms, ILS-DLA, for sparse signal representation. Digital Signal Process. **17**(1), 32–49 (2007)
18. K. Skretting, K. Engan, Recursive least squares dictionary learning algorithm. IEEE Trans. Signal Process. **58**(4), 2121–2130 (2010)
19. S.K. Sahoo, A. Makur, Dictionary training for sparse representation as generalization of k-means clustering. IEEE Signal Process. Lett. **20**(6), 587–590 (2013)
20. C.A. Metzler, A. Maleki, R.G. Baraniuk, From denoising to compressed sensing. IEEE Trans. Inf. Theory **62**(9), 5117–5144 (2016)
21. L. Anitori, A. Maleki, M. Otten, R.G. Baraniuk, P. Hoogeboom, Design and analysis of compressed sensing radar detectors. IEEE Trans. Signal Process. **61**(4), 813–827 (2013)
22. CIV Test Images, available at: http://www.io.csic.es/PagsPers/JPortilla/image-processing/bls-gsm/63-test-images
23. Z. Li, H. Huang, S. Misra, Compressed sensing via dictionary learning and approximate message passing for multimedia internet of things. IEEE Internet Things J. **4**(2), 505–512 (2016)

Chapter 3
Turbo-Type Algorithm for Affine Rank Minimization

3.1 Problem Description

We consider a rank-r matrix $\boldsymbol{X}_0 \in \mathbb{R}^{n_1 \times n_2}$ with the integers r, n_1, and n_2 satisfying $r \ll \min(n_1, n_2)$. We aim to recover $\boldsymbol{X}_0$ from an affine measurement given by

$$\boldsymbol{y} = \mathcal{A}(\boldsymbol{X}_0) \in \mathbb{R}^m \tag{3.1}$$

where $\mathcal{A} : \mathbb{R}^{n_1 \times n_2} \to \mathbb{R}^m$ is a linear map with $m < n_1 n_2 = n$. When $\mathcal{A}$ is a general linear operator such as Gaussian operators and partial orthogonal operators, we refer to the problem as *low-rank matrix recovery*; when $\mathcal{A}$ is a selector that outputs a subset of the entries of $\boldsymbol{X}_0$, we refer to the problem as *matrix completion*.

In real-world applications, perfect measurements are rare, and noise is naturally introduced in the measurement process. That is, we want to recover $\boldsymbol{X}_0$ from a noisy measurement of

$$\boldsymbol{y} = \mathcal{A}(\boldsymbol{X}_0) + \boldsymbol{n} \tag{3.2}$$

where $\boldsymbol{n} \in \mathbb{R}^m$ is a Gaussian noise with zero mean and covariance $\sigma^2 \boldsymbol{I}$ and is independent of $\mathcal{A}(\boldsymbol{X}_0)$. To recover the low-rank matrix $\boldsymbol{X}_0$ from (3.2), we turn to the following formulation of the stable ARM problem:

$$\begin{aligned} &\min_{\boldsymbol{X}} \| \boldsymbol{y} - \mathcal{A}(\boldsymbol{X}) \|_2^2 \\ &\text{s.t. } \operatorname{rank}(\boldsymbol{X}) \le r. \end{aligned} \tag{3.3}$$

The problem in (3.3) is still NP-hard and difficult to solve. Several suboptimal algorithms have been proposed to yield approximate solutions to (3.3). For example, the author in [1] proposed an alternating minimization method to factorize rank-

X. Yuan and Z. Xue, *Turbo Message Passing Algorithms for Structured Signal Recovery*, SpringerBriefs in Computer Science,
https://doi.org/10.1007/978-3-030-54762-2_3

r matrix X_0 as the product of two matrices with dimension $n_1 \times r$ and $r \times n_2$, respectively. This method is more efficient in storage than SDP and SVT methods, especially when large-dimension low-rank matrices are involved. A second approach borrows the idea of iterative hard thresholding (IHT) for compressed sensing. For example, the singular value projection (SVP) algorithm [2] for stable ARM can be viewed as a counterpart of the IHT algorithm [3] for compressed sensing. SVP solves the stable ARM problem by combining the projected gradient method with singular value decomposition (SVD). An improved version of SVP, termed normalized IHT (NIHT) [4], adaptively selects the step size of the gradient descent step of SVP, rather than using a fixed step size. These algorithms involve a projection step which projects a matrix into a low-rank space using truncated SVD. In [5], a Riemannian method, termed RGrad, was proposed to extend NIHT by projecting the search direction of gradient descent into a low dimensional space. Compared with the alternating minimization method, these IHT-based algorithms exhibit better convergence performance with lower computational complexity. Furthermore, the convergence of these IHT-based algorithms is guaranteed when a certain restricted isometry property (RIP) holds [2, 4, 5].

In this chapter, we aim to design low-complexity iterative algorithms to solve the stable ARM problem based on the turbo message passing principles, a different perspective from the existing approaches mentioned above. Specifically, we present a turbo-type algorithm, termed turbo-type affine rank minimization (TARM), for solving the stable ARM problem, as inspired by the turbo compressed sensing (Turbo-CS) algorithm for sparse signal recovery [6, 7].

3.2 The TARM Algorithm

As inspired by the success of the Turbo-CS algorithm for sparse signal recovery, we borrow the idea of turbo message passing and present the TARM algorithm for the affine rank minimization problem in this section.

3.2.1 Algorithm Description

The diagram of TARM is illustrated in Fig. 3.1 and the detailed steps of TARM are presented in Algorithm 3. We use index t to denote the t-th iteration. There are two concatenated modules in TARM:

1. Module A:

    - Step 1: We estimate the low-rank matrix X_0 via a linear estimator $\mathcal{E}(\cdot)$ based on the observation y and the input $X^{(t-1)}$:

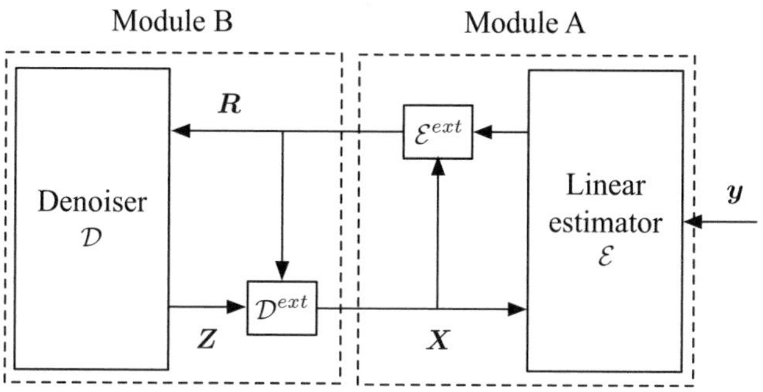

Fig. 3.1 The diagram of the TARM algorithm

$$\mathcal{E}(\boldsymbol{X}^{(t-1)}) = \boldsymbol{X}^{(t-1)} + \gamma_t \mathcal{A}^T(\boldsymbol{y} - \mathcal{A}(\boldsymbol{X}^{(t-1)})) \tag{3.4}$$

where γ_t is a certain given coefficient.

- Step 2: The extrinsic estimate of $\boldsymbol{X}_0$ is then given by

$$\begin{aligned} \boldsymbol{R}^{(t)} &= \mathcal{E}^{ext}(\boldsymbol{X}^{(t-1)}, \mathcal{E}(\boldsymbol{X}^{(t-1)})) \\ &= \boldsymbol{X}^{(t-1)} + \frac{\mu_t}{\gamma_t}(\mathcal{E}(\boldsymbol{X}^{(t-1)}) - \boldsymbol{X}^{(t-1)}) \\ &= \boldsymbol{X}^{(t-1)} + \mu_t \mathcal{A}^T(\boldsymbol{y} - \mathcal{A}(\boldsymbol{X}^{(t-1)})) \end{aligned} \tag{3.5}$$

where $\mathcal{E}^{ext}(\boldsymbol{X}^{(t-1)}, \mathcal{E}(\boldsymbol{X}^{(t-1)}))$ linearly combines the inputs $\boldsymbol{X}^{(t-1)}$ and $\mathcal{E}(\boldsymbol{X}^{(t-1)})$ with μ_t being a known coefficient.[1]

We combine (3.4) and (3.5) into a linear estimation (Line 3 of Algorithm 3) since both operations are linear.

2. Module B:

- Step 1: The output of Module A, i.e., $\boldsymbol{R}^{(t)}$, is passed to a denoiser $\mathcal{D}(\cdot)$ which suppresses the estimation error by exploiting the low-rank structure of $\boldsymbol{X}_0$ (Line 4 of Algorithm 3).
- Step 2: The denoised output $\boldsymbol{Z}^{(t)}$ is passed to a linear function $\mathcal{D}^{ext}(\cdot, \cdot)$ which linearly combines $\boldsymbol{Z}^{(t)}$ and $\boldsymbol{R}^{(t)}$ (Line 5 of Algorithm 3).

Denoiser $\mathcal{D}(\cdot)$ can be chosen as the best rank-r approximation [8] or the singular value thresholding (SVT) denoiser [9]. We focus on the best rank-r approximation defined by

[1] From the turbo principle, μ_t is chosen to ensure that the output error of Module A is uncorrelated with the input error, i.e. $\langle \boldsymbol{R}^{(t)} - \boldsymbol{X}_0, \boldsymbol{X}^{(t-1)} - \boldsymbol{X}_0 \rangle = 0$. More detailed discussions can be found in Sect. 3.2.2.

Algorithm 3 TARM for affine rank minimization

Input: $\mathcal{A}$, $\mathbf{y}$, $\mathbf{X}^{(0)} = \mathbf{0}$, $t = 0$
1: **while** the stopping criterion is not met **do**
2: $t = t + 1$
3: $\mathbf{R}^{(t)} = \mathbf{X}^{(t-1)} + \mu_t \mathcal{A}^T(\mathbf{y} - \mathcal{A}(\mathbf{X}^{(t-1)}))$
4: $\mathbf{Z}^{(t)} = \mathcal{D}(\mathbf{R}^{(t)})$
5: $\mathbf{X}^{(t)} = \mathcal{D}^{ext}(\mathbf{R}^{(t)}, \mathbf{Z}^{(t)}) = c_t(\mathbf{Z}^{(t)} - \alpha_t \mathbf{R}^{(t)})$
6: **end while**
Output: $\mathbf{Z}^{(t)}$

$$\mathcal{D}(\mathbf{R}) = \sum_{i=1}^{r} \sigma_i \mathbf{u}_i \mathbf{v}_i^T \tag{3.6}$$

where σ_i, $\mathbf{u}_i$, and $\mathbf{v}_i$ are, respectively, the i-th singular value and the corresponding left and right singular vectors of the input $\mathbf{R}$.

In the above, the superscript "ext" stands for *extrinsic message*. Step 2 of each module is dedicated to the calculation of extrinsic messages which is a major difference of TARM from its counterpart algorithms. In particular, we note that in TARM when $c_t = 1$ and $\alpha_t = 0$ for any t, the algorithm reduces to the SVP or NIHT algorithm (depending on the choice of μ_t). As such, the key difference of TARM from SVP and NITH resides in the choice of these parameters. By optimizing these parameters, the TARM algorithm aims to find a better descent direction in each iteration, so as to achieve a convergence rate much higher than SVP and NIHT.

3.2.2 Determining the Parameters of TARM

In this subsection, we discuss how to determine the parameters $\{\mu_t\}$, $\{c_t\}$, and $\{\alpha_t\}$ based on turbo message passing. Turbo message passing was first applied to iterative decoding of turbo codes and then extended for solving compressed sensing problems as studied in Chap. 2. Following the turbo message passing rule in Chap. 2, three conditions for the calculation of extrinsic messages are presented:

- Condition 1:

$$\left\langle \mathbf{R}^{(t)} - \mathbf{X}_0, \mathbf{X}^{(t-1)} - \mathbf{X}_0 \right\rangle = 0; \tag{3.7}$$

- Condition 2:

$$\left\langle \mathbf{R}^{(t)} - \mathbf{X}_0, \mathbf{X}^{(t)} - \mathbf{X}_0 \right\rangle = 0; \tag{3.8}$$

- Condition 3: For given $\mathbf{X}^{(t-1)}$,

$$\|\boldsymbol{X}^{(t)} - \boldsymbol{X}_0\|_F^2 \text{ is minimized under (3.7) and (3.8).} \tag{3.9}$$

In the above, Conditions 1 and 2 follow from [7, Eq. 7, Eq. 14], and Condition 3 follows from second condition in Chap. 2. Condition 1 ensures that the input and output estimation errors of Module A are uncorrelated. Similarly, Condition 2 ensures that the input and output estimation errors of Module B are uncorrelated. Condition 3 ensures that the output estimation error of Module B is minimized over $\{\mu_t, c_t, \alpha_t\}$ for each iteration t.

We have the following lemma for $\{\mu_t, c_t, \alpha_t\}$, with the proof given in Appendix 1.

Lemma 3.1 *If Conditions 1–3 hold, then*

$$\mu_t = \frac{\|\boldsymbol{X}^{(t-1)} - \boldsymbol{X}_0\|_F^2}{\left\langle \mathcal{A}(\boldsymbol{X}^{(t-1)} - \boldsymbol{X}_0) - \boldsymbol{n}, \mathcal{A}(\boldsymbol{X}^{(t-1)} - \boldsymbol{X}_0)\right\rangle} \tag{3.10a}$$

$$\alpha_t = \frac{-b_t \pm \sqrt{b_t^2 - 4a_t d_t}}{2a_t} \tag{3.10b}$$

$$c_t = \frac{\left\langle \boldsymbol{Z}^{(t)} - \alpha_t \boldsymbol{R}^{(t)}, \boldsymbol{R}^{(t)}\right\rangle}{\|\boldsymbol{Z}^{(t)} - \alpha_t \boldsymbol{R}^{(t)}\|_F^2}, \tag{3.10c}$$

with

$$a_t = \|\boldsymbol{R}^{(t)}\|_F^2 \|\boldsymbol{R}^{(t)} - \boldsymbol{X}_0\|_F^2 \tag{3.11a}$$

$$b_t = -\|\boldsymbol{R}^{(t)}\|_F^2 \left\langle \boldsymbol{R}^{(t)} - \boldsymbol{X}_0, \boldsymbol{Z}^{(t)}\right\rangle - \|\boldsymbol{Z}^{(t)}\|_F^2 \|\boldsymbol{R}^{(t)} - \boldsymbol{X}_0\|_F^2 + \|\boldsymbol{Z}^{(t)}\|_F^2 \left\langle \boldsymbol{R}^{(t)} - \boldsymbol{X}_0, \boldsymbol{X}_0\right\rangle \tag{3.11b}$$

$$d_t = \|\boldsymbol{Z}^{(t)}\|_F^2 \left\langle \boldsymbol{R}^{(t)} - \boldsymbol{X}_0, \boldsymbol{Z}^{(t)} - \boldsymbol{X}_0\right\rangle. \tag{3.11c}$$

Remark 3.1 In (3.10b), α_t has two possible choices and only one of them minimizes the error in (3.9). From the discussion below (3.35), minimizing the square error in (3.9) is equivalent to minimizing $\|\boldsymbol{X}^{(t)} - \boldsymbol{R}^{(t)}\|_F^2$. We have

$$\left\|\boldsymbol{X}^{(t)} - \boldsymbol{R}^{(t)}\right\|_F^2$$

$$= \left\|c_t(\boldsymbol{Z}^{(t)} - \alpha_t \boldsymbol{R}^{(t)}) - \boldsymbol{R}^{(t)}\right\|_F^2 \tag{3.12a}$$

$$= -\frac{\left\langle \boldsymbol{Z}^{(t)} - \alpha_t \boldsymbol{R}^{(t)}, \boldsymbol{R}^{(t)}\right\rangle^2}{\|\boldsymbol{Z}^{(t)} - \alpha_t \boldsymbol{R}^{(t)}\|_F^2} + \|\boldsymbol{R}^{(t)}\|_F^2 \tag{3.12b}$$

where (3.12a) follows from substituting $\boldsymbol{X}^{(t)}$ in Line 5 of Algorithm 3, and (3.12b) follows by substituting c_t in (3.10c). Since $\|\boldsymbol{R}^{(t)}\|_F^2$ is invariant to α_t, minimizing $\|\boldsymbol{X}^{(t)} - \boldsymbol{R}^{(t)}\|_F^2$ is equivalent to maximizing $\frac{\langle \boldsymbol{Z}^{(t)} - \alpha_t \boldsymbol{R}^{(t)}, \boldsymbol{R}^{(t)} \rangle^2}{\|\boldsymbol{Z}^{(t)} - \alpha_t \boldsymbol{R}^{(t)}\|_F^2}$. We choose α_t that gives a larger value of $\frac{\langle \boldsymbol{Z}^{(t)} - \alpha_t \boldsymbol{R}^{(t)}, \boldsymbol{R}^{(t)} \rangle^2}{\|\boldsymbol{Z}^{(t)} - \alpha_t \boldsymbol{R}^{(t)}\|_F^2}$.

Remark 3.2 Similarly to SVP [2] and NIHT [4], the convergence of TARM can be analyzed by assuming that the linear operator $\mathcal{A}$ satisfies the RIP. The convergence rate of TARM is much faster than those of NIHT and SVP (provided that $\{\alpha_t\}$ are sufficiently small). More detailed discussions are presented in Appendix 2.

We emphasize that the parameters μ_t, α_t, and c_t in (3.10) are actually difficult to evaluate since $\boldsymbol{X}_0$ and $\boldsymbol{n}$ are unknown. This means that Algorithm 3 cannot rely on (3.10) to determine μ_t, α_t, and c_t. In the following, we focus on how to approximately evaluate these parameters to yield practical algorithms. Based on different choices of the linear operator $\mathcal{A}$, our discussions are divided into two parts, namely, low-rank matrix recovery and matrix completion.

3.3 Low-Rank Matrix Recovery

3.3.1 Preliminaries

In this section, we consider recovering $\boldsymbol{X}_0$ from measurement in (3.2) when the linear operator $\mathcal{A}$ is right-orthogonally invariant and $\boldsymbol{X}_0$ is generated by following the random models described in [10].[2] Denote the vector form of an arbitrary matrix $\boldsymbol{X} \in \mathbb{R}^{n_1 \times n_2}$ by $\boldsymbol{x} = \text{vec}(\boldsymbol{X}) = [\boldsymbol{x}_1^T, \boldsymbol{x}_2^T, \ldots, \boldsymbol{x}_n^T]^T$, where $\boldsymbol{x}_i$ is the ith column of $\boldsymbol{X}$. The linear operator $\mathcal{A}$ can be generally expressed as $\mathcal{A}(\boldsymbol{X}) = \boldsymbol{A}\text{vec}(\boldsymbol{X}) = \boldsymbol{A}\boldsymbol{x}$ where $\boldsymbol{A} \in \mathbb{R}^{m \times n}$ is a matrix representation of $\mathcal{A}$. The adjoint operator $\mathcal{A}^T : \mathbb{R}^m \rightarrow \mathbb{R}^{n_1 \times n_2}$ is defined by the transpose of $\boldsymbol{A}$ with $\boldsymbol{x}' = \text{vec}(\boldsymbol{X}') = \text{vec}(\mathcal{A}^T(\boldsymbol{y}')) = \boldsymbol{A}^T \boldsymbol{y}'$. Consider a linear operator $\mathcal{A}$ with matrix form $\boldsymbol{A}$, the SVD of $\boldsymbol{A}$ is $\boldsymbol{A} = \boldsymbol{U}_A \boldsymbol{\Sigma}_A \boldsymbol{V}_A^T$, where $\boldsymbol{U}_A \in \mathbb{R}^{m \times m}$ and $\boldsymbol{V}_A \in \mathbb{R}^{n \times n}$ are orthogonal matrices and $\boldsymbol{\Sigma}_A \in \mathbb{R}^{m \times n}$ is a diagonal matrix.

Definition 3.1 If $\boldsymbol{V}_A$ is a Haar distributed random matrix [11] independent of $\boldsymbol{\Sigma}_A$, we say that $\mathcal{A}$ is a right-orthogonally invariant linear (ROIL) operator.[3]

[2]The generation models of $\boldsymbol{X}_0$ can be found in the definitions before Assumption 2.4 in [10]. This choice allows us to borrow the results of [10] in our analysis; see (3.59).

[3]In this section, we focus on ROIL operators so that the algorithm parameters can be determined by following the discussion in Appendix 2. However, we emphasize that the proposed TARM algorithm applies to low-rank matrix recovery even when $\mathcal{A}$ is not a ROIL operator. In this case, the only difference is that the algorithm parameters shall be determined by following the heuristic methods described in Sect. 3.2.2.

We focus on two types of ROIL operators: partial orthogonal ROIL operators where the matrix form of $\mathcal{A}$ satisfies $\boldsymbol{A}\boldsymbol{A}^T = \boldsymbol{I}$, and Gaussian ROIL operators where the elements of $\boldsymbol{A}$ are i.i.d. Gaussian with zero mean. For convenience of discussion, the linear operator $\mathcal{A}$ is normalized such that the l_2-norm of each row of $\boldsymbol{A}$ is 1. It is worth noting that from the perspective of the algorithm, $\boldsymbol{A}$ is deterministic since $\boldsymbol{A}$ is known by the algorithm. However, the randomness of $\boldsymbol{A}$ has impact on parameter design and performance analysis, as detailed in what follows.

We now present two assumptions that are useful in determining the algorithm parameters in the following subsection.

Assumption 3.1 For each iteration t, Module A's input estimation error $\boldsymbol{X}^{(t-1)} - \boldsymbol{X}_0$ is independent of the orthogonal matrix $\boldsymbol{V}_A$ and the measurement noise $\boldsymbol{n}$.

Assumption 3.2 For each iteration t, the output error of Module A, given by $\boldsymbol{R}^{(t)} - \boldsymbol{X}_0$, is an i.i.d. Gaussian noise, i.e., the elements of $\boldsymbol{R}^{(t)} - \boldsymbol{X}_0$ are independently and identically drawn from $\mathcal{N}(0, v_t)$, where v_t is the output variance of Module A at iteration t.

The above two assumptions will be verified for ROIL operators by the numerical results presented in Appendix 4. Similar assumptions have been introduced in the design of Turbo-CS in [6] (see also [12]). Later, these assumptions were rigorously analyzed in [13, 14] using the conditioning technique [15]. Based on that, state evolution was established to characterize the behavior of the Turbo-CS algorithm.

Assumptions 3.1 and 3.2 allow to decouple Module A and Module B in the analysis of the TARM algorithm. We will derive two mean square error (MSE) transfer functions, one for each module, to characterize the behavior of the TARM algorithm. The details will be presented in Appendix 3.

3.3.2 Parameter Design

We now determine the parameters in (3.10) when ROIL operators are involved. We show that (3.10) can be approximately evaluated without the knowledge of $\boldsymbol{X}_0$. Since $\{c_t\}$ in (3.10c) can be readily computed given $\{\alpha_t\}$, we focus on the calculation of $\{\mu_t\}$ and $\{\alpha_t\}$.

We start with μ_t. From (3.10a), we have

$$\mu_t = \frac{\|\boldsymbol{X}^{(t-1)} - \boldsymbol{X}_0\|_F^2}{\langle \mathcal{A}(\boldsymbol{X}^{(t-1)} - \boldsymbol{X}_0) - \boldsymbol{n}, \mathcal{A}(\boldsymbol{X}^{(t-1)} - \boldsymbol{X}_0)\rangle} \tag{3.13a}$$

$$\approx \frac{\|\boldsymbol{X}^{(t-1)} - \boldsymbol{X}_0\|_F^2}{\|\mathcal{A}(\boldsymbol{X}^{(t-1)} - \boldsymbol{X}_0)\|_2^2} \tag{3.13b}$$

$$= \frac{1}{\tilde{\boldsymbol{x}}^T \boldsymbol{V}_A \boldsymbol{\Sigma}_A^T \boldsymbol{\Sigma}_A \boldsymbol{V}_A^T \tilde{\boldsymbol{x}}} \tag{3.13c}$$

$$= \frac{1}{\boldsymbol{v}_A^T \boldsymbol{\Sigma}_A^T \boldsymbol{\Sigma}_A \boldsymbol{v}_A} \approx \frac{n}{m} \tag{3.13d}$$

where (3.13b) holds approximately for a relatively large matrix size since $\left\langle \boldsymbol{n}, \mathcal{A}(\boldsymbol{X}^{(t-1)} - \boldsymbol{X}_0) \right\rangle \approx 0$ from Assumption 3.1, (3.13c) follows by utilizing the matrix form of $\mathcal{A}$ and $\tilde{\boldsymbol{x}} = \frac{\mathrm{vec}(\boldsymbol{X}^{(t-1)} - \boldsymbol{X}_0)}{\|\boldsymbol{X}^{(t-1)} - \boldsymbol{X}_0\|_F}$, and (3.13d) follows by letting $\boldsymbol{v}_A = \boldsymbol{V}_A^T \tilde{\boldsymbol{x}}$. $\boldsymbol{V}_A$ is Haar distributed and from Assumption 3.1 is independent of $\tilde{\boldsymbol{x}}$, implying that $\boldsymbol{v}_A$ is a unit vector uniformly distributed over the sphere $\|\boldsymbol{v}_A\|_2 = 1$. Then, the approximation in (3.13d) follows by noting $\mathrm{Tr}(\boldsymbol{\Sigma}_A^T \boldsymbol{\Sigma}_A) = m$.

We next consider the approximation of α_t. We first note

$$\left\langle \boldsymbol{R}^{(t)} - \boldsymbol{X}_0, \boldsymbol{X}^{(t)} - \boldsymbol{X}_0 \right\rangle \tag{3.14a}$$

$$= \left\langle \boldsymbol{R}^{(t)} - \boldsymbol{X}_0, c_t(\boldsymbol{Z}^{(t)} - \alpha_t \boldsymbol{R}^{(t)}) - \boldsymbol{X}_0 \right\rangle \tag{3.14b}$$

$$\approx c_t \left\langle \boldsymbol{R}^{(t)} - \boldsymbol{X}_0, \boldsymbol{Z}^{(t)} - \alpha_t \boldsymbol{R}^{(t)} \right\rangle \tag{3.14c}$$

where (3.14a) follows by substituting $\boldsymbol{X}^{(t)}$ in line 5 of Algorithm 3, and (3.14b) follows from $\left\langle \boldsymbol{R}^{(t)} - \boldsymbol{X}_0, \boldsymbol{X}_0 \right\rangle \approx 0$ (implying that the error $\boldsymbol{R}^{(t)} - \boldsymbol{X}_0$ is uncorrelated with the original signal $\boldsymbol{X}_0$). Combining (3.14) and Condition 2 in (3.8), we have

$$\alpha_t = \frac{\left\langle \boldsymbol{R}^{(t)} - \boldsymbol{X}_0, \boldsymbol{Z}^{(t)} \right\rangle}{\left\langle \boldsymbol{R}^{(t)} - \boldsymbol{X}_0, \boldsymbol{R}^{(t)} \right\rangle} \tag{3.15a}$$

$$\approx \frac{\left\langle \boldsymbol{R}^{(t)} - \boldsymbol{X}_0, \mathcal{D}(\boldsymbol{R}^{(t)}) \right\rangle}{\left\langle \boldsymbol{R}^{(t)} - \boldsymbol{X}_0, \boldsymbol{R}^{(t)} - \boldsymbol{X}_0 \right\rangle} \tag{3.15b}$$

$$\approx \frac{\left\langle \boldsymbol{R}^{(t)} - \boldsymbol{X}_0, \mathcal{D}(\boldsymbol{R}^{(t)}) \right\rangle}{n v_t} \tag{3.15c}$$

$$\approx \frac{1}{n} \sum_{i,j} \frac{\partial \mathcal{D}(\boldsymbol{R}^{(t)})_{i,j}}{\partial R_{i,j}^{(t)}} = \frac{1}{n} \mathrm{div}(\mathcal{D}(\boldsymbol{R}^{(t)})) \tag{3.15d}$$

where (3.15b) follows from $\boldsymbol{Z}^{(t)} = \mathcal{D}(\boldsymbol{R}^{(t)})$ and $\left\langle \boldsymbol{R}^{(t)} - \boldsymbol{X}_0, \boldsymbol{X}_0 \right\rangle \approx 0$, (3.15c) follows from Assumption 3.2 that the elements of $\boldsymbol{R}^{(t)} - \boldsymbol{X}_0$ are i.i.d. Gaussian with zero mean and variance v_t, (3.15d) follows from Stein's lemma [16] since we approximate the entries of $\boldsymbol{R}^{(t)} - \boldsymbol{X}_0$ as i.i.d. Gaussian distributed.

3.3.3 State Evolution

We now characterize the performance of TARM for low-rank matrix recovery based on Assumptions 3.1 and 3.2.

We first consider the MSE behavior of Module A. Denote the output MSE of Module A at iteration t by

$$\mathrm{MSE}_A^{(t)} = \frac{1}{n}\|\boldsymbol{R}^{(t)} - \boldsymbol{X}_0\|_F^2. \tag{3.16}$$

The following theorem gives the asymptotic MSE of Module A when the dimension of $\boldsymbol{X}_0$ goes to infinity, with the proof given in Appendix 3.

Theorem 3.1 *Assume that Assumption 3.1 holds, and let* $\mu = \frac{n}{m}$. *Then,*

$$\mathrm{MSE}_A^{(t)} \xrightarrow{a.s.} f(\tau_t) \tag{3.17}$$

as $m, n \to \infty$ *with* $\frac{m}{n} \to \delta$, *where* $\frac{1}{n}\|\boldsymbol{X}^{(t-1)} - \boldsymbol{X}_0\|_F^2 \to \tau_t$ *as* $n \to \infty$. *For partial orthogonal ROIL operator* $\mathcal{A}$,

$$f(\tau) = (1/\delta - 1)\tau + 1/\delta\sigma^2 \tag{3.18a}$$

and for Gaussian ROIL operator $\mathcal{A}$,

$$f(\tau) = 1/\delta\tau + 1/\delta\sigma^2. \tag{3.18b}$$

We now consider the MSE behavior of Module B. We start with the following useful lemma, with the proof given in Appendix 4.

Lemma 3.2 *Assume that* $\boldsymbol{R}^{(t)}$ *satisfies Assumption 3.2,* $\|\boldsymbol{X}_0\|_F^2 = n$, *and the empirical distribution of eigenvalue* θ *of* $\frac{1}{n_2}\boldsymbol{X}_0^T\boldsymbol{X}_0$ *converges almost surely to the density function* $p(\theta)$ *as* $n_1, n_2, r \to \infty$ *with* $\frac{n_1}{n_2} \to \rho$, $\frac{r}{n_2} \to \lambda$. *Then,*

$$\alpha_t \xrightarrow{a.s.} \alpha(v_t) \tag{3.19a}$$

$$c_t \xrightarrow{a.s.} c(v_t) \tag{3.19b}$$

as $n_1, n_2, r \to \infty$ *with* $\frac{n_1}{n_2} \to \rho$, $\frac{r}{n_2} \to \lambda$, *where*

$$\alpha(v) = \left|1 - \frac{1}{\rho}\right|\lambda + \frac{1}{\rho}\lambda^2 + 2\left(\min\left(1, \frac{1}{\rho}\right) - \frac{\lambda}{\rho}\right)\lambda\Delta_1(v) \tag{3.20a}$$

$$c(v) = \frac{1 + \lambda(1 + \frac{1}{\rho})v + \lambda v^2\Delta_2 - \alpha(v)(1 + v)}{(1 - 2\alpha(v))(1 + \lambda(1 + \frac{1}{\rho})v + \lambda v^2\Delta_2) + \alpha(v)^2(1 + v)} \tag{3.20b}$$

with Δ_1 and Δ_2 defined by

$$\Delta_1(v) = \int_0^\infty \frac{(v+\theta^2)(\rho v+\theta^2)}{(\sqrt{\rho}v-\theta^2)^2} p(\theta)d\theta \tag{3.21a}$$

$$\Delta_2 = \int_0^\infty \frac{1}{\theta^2} p(\theta)d\theta. \tag{3.21b}$$

Denote the output MSE of Module B at iteration t by

$$\mathrm{MSE}_B^{(t)} = \frac{1}{n}\|\boldsymbol{X}^{(t)} - \boldsymbol{X}_0\|_F^2. \tag{3.22}$$

The output MSE of Module B is characterized by the following theorem.

Theorem 3.2 *Assume that Assumption 3.2 holds, and let $\|\boldsymbol{X}_0\|_F^2 = n$. Then, the output MSE of Module B*

$$\mathrm{MSE}_B^{(t)} \xrightarrow{a.s.} g(v_t) \tag{3.23}$$

as $n_1, n_2, r \to \infty$ with $\frac{n_1}{n_2} \to \rho$, $\frac{r}{n_2} \to \lambda$, where

$$g(v_t) \triangleq \frac{v_t - \lambda\left(1+\frac{1}{\rho}\right)v_t - \lambda v_t^2 \Delta_2}{\frac{v_t-\lambda\left(1+\frac{1}{\rho}\right)v_t-\lambda v_t^2\Delta_2}{1+\lambda\left(1+\frac{1}{\rho}\right)v_t+\lambda v_t^2\Delta_2}\alpha(v_t)^2 + (1-\alpha(v_t))^2} - v_t \tag{3.24}$$

α and Δ_2 are given in Lemma 3.2, and $\frac{1}{n}\|\boldsymbol{R}^{(t)} - \boldsymbol{X}_0\|_F^2 \xrightarrow{a.s.} v_t$.

Remark 3.3 Δ_1 and Δ_2 in (3.21) may be difficult to obtain since $p(\theta)$ is usually unknown in practical scenarios. We now introduce an approximate MSE expression that does not depend on $p(\theta)$:

$$g(v_t) \approx \bar{g}(v_t) \triangleq \frac{v_t - \lambda(1+\frac{1}{\rho})v_t}{(1-\alpha)^2} - v_t \tag{3.25}$$

where $\alpha = \alpha(0) = \left|1-\frac{1}{\rho}\right|\lambda - \frac{1}{\rho}\lambda^2 + 2\min\left(1, \frac{1}{\rho}\right)\lambda$. Compared with $g(v_t)$, $\bar{g}(v_t)$ omits two terms $-\lambda v_t^2 \Delta$ and $\frac{v_t-\lambda\left(1+\frac{1}{\rho}\right)v_t-\lambda v_t^2\Delta}{1+\lambda\left(1+\frac{1}{\rho}\right)v_t+\lambda v_t^2\Delta}\alpha(v_t)^2$ and replaces $\alpha(v_t)$ by α. Recall that v_t is the mean square error at the t-iteration. As the iteration proceeds, we have $v_t \ll 1$, and hence $g(v_t)$ can be well approximated by $\bar{g}(v_t)$, as seen later from Fig. 3.3.

Combining Theorems 3.1 and 3.2, we can characterize the MSE evolution of TARM by

$$v_t = f(\tau_t) \tag{3.26a}$$

$$\tau_{t+1} = g(v_t). \tag{3.26b}$$

The fixed point of TARM's MSE evolution in (3.26) is given by

$$\tau^* = g(f(\tau^*)). \tag{3.27}$$

The above fixed point equation can be used to analyze the phase transition curves of the TARM algorithm. It is clear that the fixed point τ^* of (3.27) is a function of $\{\delta, \rho, \lambda, \Delta, \sigma\}$. For any given $\{\delta, \rho, \lambda, \Delta, \sigma\}$, we say that the TARM algorithm is successful if the corresponding τ^* is below a certain predetermined threshold. The critical values of $\{\delta, \rho, \lambda, \Delta, \sigma\}$ define the phase transition curves of the TARM algorithm.

3.3.4 *Numerical Results*

Simulation settings are as follows. For the case of partial orthogonal ROIL operators, we generate a partial orthogonal ROIL operator with the matrix form

$$\boldsymbol{A} = \boldsymbol{S}\boldsymbol{W}\boldsymbol{\Theta} \tag{3.28}$$

where $\boldsymbol{S} \in \mathbb{R}^{m\times n}$ is a random selection matrix, $\boldsymbol{W} \in \mathbb{R}^{n\times n}$ is a discrete cosine transform (DCT) matrix, and $\boldsymbol{\Theta}$ is a diagonal matrix with diagonal entries being 1 or -1 randomly. For the case of Gaussian ROIL operators, we generate an i.i.d. Gaussian random matrix of size $m \times n$ with elements drawn from $\mathcal{N}(0, \frac{1}{n})$. The rank-$r$ matrix $\boldsymbol{X}_0 \in \mathbb{R}^{n_1\times n_2}$ is generated by the product of two i.i.d. Gaussian matrices of size $n_1 \times r$ and $r \times n_2$.

3.3.4.1 Verification of the Assumptions

We first verify Assumption 3.1 using Table 3.1. Recall that if Assumption 3.1 holds, the approximations in the calculation of μ_t in (3.13) become accurate. Thus, we compare the value of μ_t calculated by (3.10a) with $\mu_t = \frac{n}{m}$ by (3.13). We record the μ_t of the first 8 iterations of TARM in Table 3.1 for low-rank matrix recovery with a partial orthogonal ROIL operator. As shown in Table 3.1, the approximation $\mu_t = \frac{n}{m}$ is close to the real value calculated by (3.10a) which serves as an evidence of the

Table 3.1 μ_t calculated by (3.10a) for the first to eighth iterations of one random realization of the algorithm with a partial orthogonal ROIL operator

Iteration t	1	2	3	4	5	6	7	8
$\frac{n}{m} = 2.5$	2.5005	2.4833	2.4735	2.4440	2.4455	2.4236	2.4465	2.4244
$\frac{n}{m} = 3.3333$	3.3474	3.3332	3.3045	3.2734	3.2512	3.2112	3.2593	3.2833
$\frac{n}{m} = 5$	5.0171	4.9500	4.9204	4.9141	4.8563	4.8102	4.7864	4.8230

$n_1 = n_2 = 1000, r = 30, \sigma = 10^{-2}$

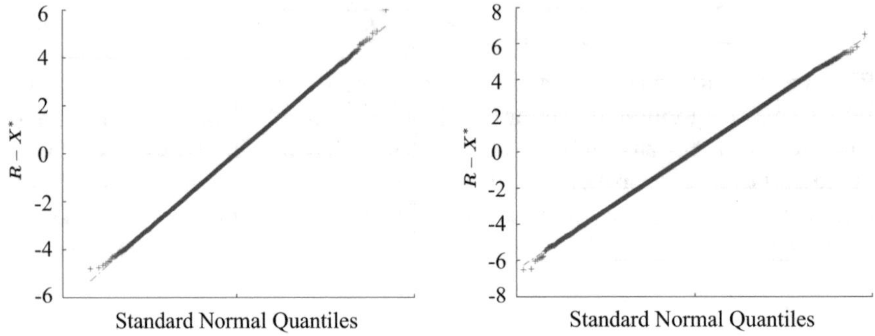

Fig. 3.2 The QQ plots of the output error of Module B in the second iteration of TARM. Left: $\mathcal{A}$ is a Gaussian ROIL operator. Right: $\mathcal{A}$ is a partial orthogonal ROIL operator. Simulation settings: $n_1 = 100, n_2 = 120, \frac{m}{n_1 n_2} = 0.3, \frac{r}{n_2} = 0.25, \sigma^2 = 0$

validity of Assumption 3.1. We then verify Assumption 3.2 using Fig. 3.2, where we plot the QQ plots of the input estimation errors of Module A with partial orthogonal and Gaussian ROIL operators. The QQ plots show that the output errors of Module A closely follow a Gaussian distribution, which agrees with Assumption 3.2.

3.3.4.2 State Evolution

We now verify the state evolution of TARM given in (3.26). We plot the simulation performance of TARM and the predicted performance by the state evolution in Fig. 3.3. From the two subfigures in Fig. 3.3, we see that the state evolution of TARM is accurate when the dimension of $\boldsymbol{X}_0$ is large enough for both partial orthogonal and Gaussian ROIL operators. We also see that the state evolution with $g(\cdot)$ replaced by the approximation in (3.25) (referred to as "Approximation" in Fig. 3.3) provides reasonably accurate performance predictions. This makes the upper bound very useful since it does not require the knowledge of the singular value distribution of $\boldsymbol{X}_0$.

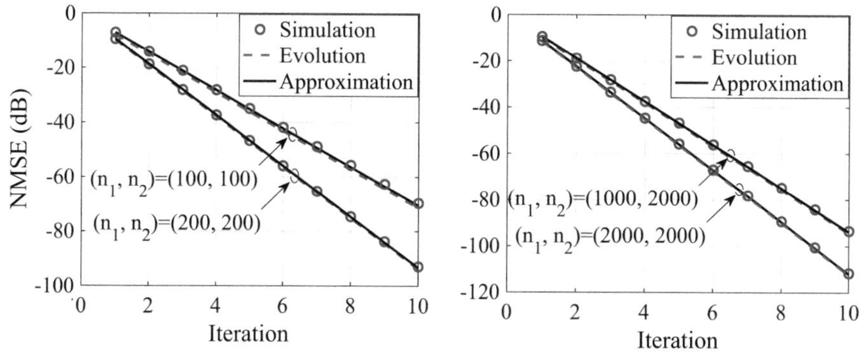

Fig. 3.3 Left: State evolution of TARM for partial orthogonal ROIL operator. $r = 40, m/n = m/(n_1 n_2) = 0.35, \sigma^2 = 0$. The size of $\boldsymbol{X}_0$ is shown in the plot. Right: State evolution of TARM for Gaussian ROIL operator. $r = 4, m/n = 0.35, \sigma^2 = 0$. The size of $\boldsymbol{X}_0$ is shown in the plot

3.3.4.3 Performance Comparisons

We compare TARM with the existing algorithms for low-rank matrix recovery problems with partial orthogonal and Gaussian ROIL operators. The following algorithms are involved: Normalized Iterative Hard Thresholding (NIHT) [4], Riemannian Gradient Descent (RGrad) [5], Riemannian Conjugate Gradient Descent (RCG) [5], and ALPS [17]. We compare these algorithms under the same settings. We plot the per iteration normalized mean square error (NMSE) defined by $\frac{\|\boldsymbol{X}^{out} - \boldsymbol{X}_0\|_F}{\|\boldsymbol{X}_0\|_F}$ where $\boldsymbol{X}^{out}$ is the output of an algorithm in Fig. 3.4. From Fig. 3.4, we see that TARM converges much faster than NIHT, RGrad, and ALPS for both Gaussian ROIL operators and partial orthogonal ROIL operators. Moreover, from the last plot in Fig. 3.4, we see TARM converges under extremely low measurement rate while the other algorithms diverge. More detailed performance comparisons are given in Sect. 3.5.

3.3.4.4 Empirical Phase Transition

The phase transition curve characterized the tradeoff between measurement rate δ and the largest rank r that an algorithm succeeds in the recovery of $\boldsymbol{X}_0$. We consider an algorithm to be successful in recovering the low-rank matrix $\boldsymbol{X}_0$ when the following conditions are satisfied: (1) the normalized mean square error $\frac{\|\boldsymbol{X}^{(t)} - \boldsymbol{X}_0\|_F^2}{\|\boldsymbol{X}_0\|_F^2} \leq 10^{-6}$; (2) the iteration number $t < 1000$. The dimension of the manifold of $n_1 \times n_2$ matrices of rank r is $r(n_1 + n_2 - r)$ [18]. Thus, for any algorithm, the minimal number of measurements for successful recovery is $r(n_1 + n_2 - r)$, i.e., $m \geq r(n_1 + n_2 - r)$. Then, an upper bound for successful recovery is $r \leq \frac{n_1 + n_2 - \sqrt{(n_1 + n_2)^2 - 4m}}{2}$. In Fig. 3.5, we plot the phase transition curves of the

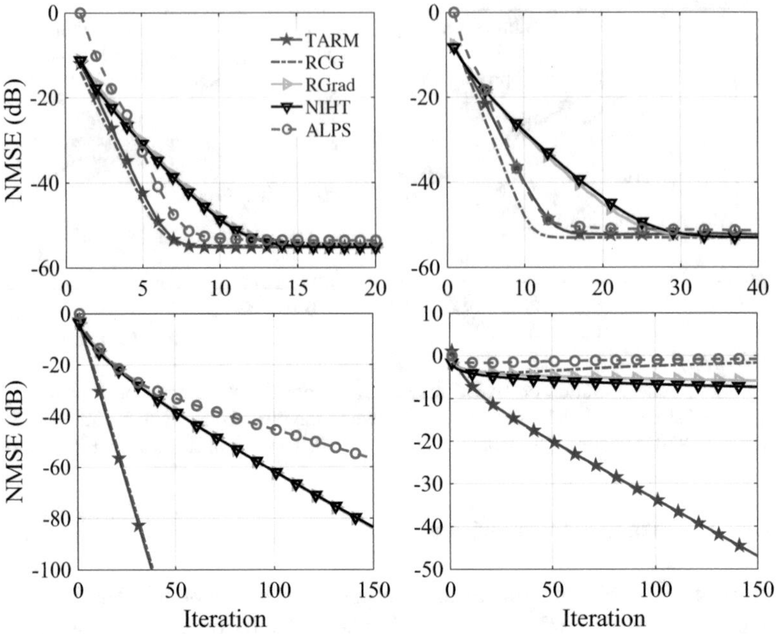

Fig. 3.4 Comparison of algorithms. Top left: $\mathcal{A}$ is a partial orthogonal ROIL operator with $n_1 = n_2 = 1000$, $r = 50$, $m/n = 0.39$, $\sigma^2 = 10^{-5}$. Top right: $\mathcal{A}$ is a Gaussian ROIL operator with $n_1 = n_2 = 80$, $r = 10$, $p = (n_1 + n_2 - r) \times r$, $m/p = 3$, $\sigma^2 = 10^{-5}$. Bottom left: $\mathcal{A}$ is a partial orthogonal ROIL operator with $n_1 = n_2 = 1000$, $r = 20$, $m/n = 0.07$, $\sigma^2 = 0$. Bottom right: $\mathcal{A}$ is a partial orthogonal ROIL operator with $n_1 = n_2 = 1000$, $r = 20$, $m/n = 0.042$, $\sigma^2 = 0$

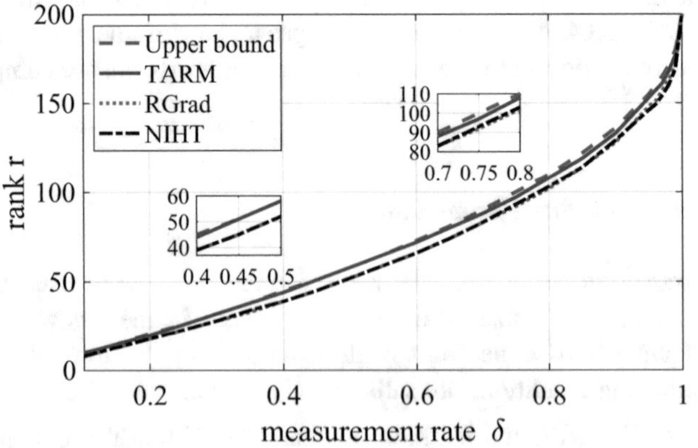

Fig. 3.5 The phase transition curves of various low-rank matrix recovery algorithms with a partial orthogonal ROIL operator. $n_1 = n_2 = 200$, $\sigma^2 = 0$. The region below each phase transition curve corresponds to the situation that the corresponding algorithm successfully recovers $\boldsymbol{X}_0$

algorithms mentioned before. From Fig. 3.5, we see that the phase transition curve of TARM is the closest to the upper bound and considerably higher than the curves of NIHT and RGrad.

3.4 Matrix Completion

In this section, we consider TARM for the matrix completion problem, where the linear operator $\mathcal{A}$ is a selector which selects a subset of the elements of the low-rank matrix $\boldsymbol{X}_0$. With such a choice of $\mathcal{A}$, the two assumptions in Sect. 3.3 for low-rank matrix recovery do not hold any more; see, e.g., Fig. 3.6. Thus, μ_t given in (3.13) and α_t in (3.15) cannot be used for matrix completion. We next discuss how to design μ_t and α_t for matrix completion.

3.4.1 Determining μ_t

The TARM algorithm is similar to SVP and NIHT as aforementioned. These three algorithms are all SVD based and a gradient descent step is involved at each iteration. The choice of descent step size μ_t is of key importance. In [4, 5], μ_t are chosen adaptively based on the idea of the steepest descent. Due to the similarity between TARM and NIHT, we follow the methods in [4, 5] and choose μ_t as

$$\mu_t = \frac{\|\mathcal{P}_S^{(t)}(\mathcal{A}^T(\boldsymbol{y} - \mathcal{A}(\boldsymbol{X}^{(t)})))\|_F^2}{\|\mathcal{A}(\mathcal{P}_S^{(t)}(\mathcal{A}^T(\boldsymbol{y} - \mathcal{A}(\boldsymbol{X}^{(t)}))))\|_2^2} \tag{3.29}$$

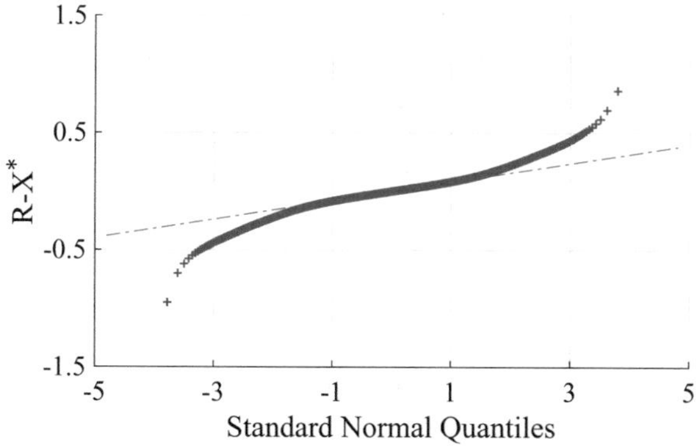

Fig. 3.6 The QQ plots of the output error of Module A in the fifth iteration of TARM for matrix completion. Simulation settings: $n_1 = 800$, $n_2 = 800$, $r = 50$, $\frac{m}{n_1 n_2} = 0.3$, $\sigma^2 = 0$

where $\mathcal{P}_{\mathcal{S}}^{(t)} : \mathbb{R}^{n_1 \times n_2} \rightarrow \mathcal{S}$ denotes a projection operator with $\mathcal{S}$ being a predetermined subspace of $\boldsymbol{R}^{n_1 \times n_2}$. The subspace $\mathcal{S}$ can be chosen as the left singular vector space of $\boldsymbol{X}^{(t)}$, the right singular vector space of $\boldsymbol{X}^{(t)}$, or the direct sum of the two subspaces [5]. Let the SVD of $\boldsymbol{X}^{(t)}$ be $\boldsymbol{X}^{(t)} = \boldsymbol{U}^{(t)}\boldsymbol{\Sigma}^{(t)}(\boldsymbol{V}^{(t)})^T$. Then, the corresponding three projection operators are given, respectively, by

$$\mathcal{P}_{\mathcal{S}_2}^{(t)}(\boldsymbol{X}) = \boldsymbol{X}\boldsymbol{V}^{(t)}(\boldsymbol{V}^{(t)})^T \tag{3.30a}$$

$$\mathcal{P}_{\mathcal{S}_1}^{(t)}(\boldsymbol{X}) = \boldsymbol{U}^{(t)}(\boldsymbol{U}^{(t)})^T\boldsymbol{X} \tag{3.30b}$$

$$\begin{aligned}\mathcal{P}_{\mathcal{S}_3}^{(t)}(\boldsymbol{X}) = {} & \boldsymbol{U}^{(t)}(\boldsymbol{U}^{(t)})^T\boldsymbol{X} + \boldsymbol{X}\boldsymbol{V}^{(t)}(\boldsymbol{V}^{(t)})^T \\ & - \boldsymbol{U}^{(t)}(\boldsymbol{U}^{(t)})^T\boldsymbol{X}\boldsymbol{V}^{(t)}(\boldsymbol{V}^{(t)})^T.\end{aligned} \tag{3.30c}$$

By combining (3.30) with (3.29), we obtain three different choices of μ_t. Later, we present numerical results to compare the impact of different choices of μ_t on the performance of TARM.

3.4.2 Determining α_t and c_t

The linear combination parameters α_t and c_t in TARM is difficult to evaluate since Assumptions 3.1 and 3.2 do not hold for TARM in the matrix completion problem. Recall that c_t is determined by α_t through (3.10c). So, we only need to determine α_t. In the following, we propose three different approaches to evaluate α_t.

The first approach is to choose α_t as in (3.15):

$$\alpha_t = \frac{\text{div}(\mathcal{D}(\boldsymbol{R}^{(t)}))}{n}. \tag{3.31}$$

We use the Monte Carlo method to compute the divergence. Specifically, the divergence of $\mathcal{D}(\boldsymbol{R}^{(t)})$ can be estimated by [19]

$$\text{div}(\mathcal{D}(\boldsymbol{R}^{(t)})) = \text{e}_{\boldsymbol{N}}\left[\left\langle \frac{\mathcal{D}(\boldsymbol{R}^{(t)} + \epsilon\boldsymbol{N}) - \mathcal{D}(\boldsymbol{R}^{(t)})}{\epsilon}, \boldsymbol{N}\right\rangle\right] \tag{3.32}$$

where $\boldsymbol{N} \in \mathbb{R}^{n_1 \times n_2}$ is a random Gaussian matrix with zero mean and unit variance entries, and ϵ is a small real number. The expectation in (3.32) can be approximated by sample mean. When the size of $\boldsymbol{R}^{(t)}$ is large, one sample is good enough for approximation. In our following simulations, we choose $\epsilon = 0.001$ and use one sample to approximate (3.32).

We now describe the second approach. Recall that we choose c_t according to (3.10c) to satisfy Condition 2: $\langle \boldsymbol{R}^{(t)} - \boldsymbol{X}_0, \boldsymbol{X}^{(t)} - \boldsymbol{X}_0 \rangle = 0$. Since $\boldsymbol{X}_0$ is unknown, finding α_t to satisfy Condition 2 is difficult. Instead, we try to find α_t that minimizes the transformed correlation of the two estimation errors:

$$\left|\left\langle \mathcal{A}(\boldsymbol{R}^{(t)} - \boldsymbol{X}_0), \mathcal{A}(\boldsymbol{X}^{(t)} - \boldsymbol{X}_0)\right\rangle\right| \tag{3.33a}$$

$$= \left|\left\langle \mathcal{A}(\boldsymbol{R}^{(t)}) - \boldsymbol{y}, \mathcal{A}(\boldsymbol{X}^{(t)}) - \boldsymbol{y}\right\rangle\right| \tag{3.33b}$$

$$= \left|\left\langle \frac{\langle \boldsymbol{Z}^{(t)} - \alpha_t \boldsymbol{R}^{(t)}, \boldsymbol{R}^{(t)}\rangle}{\|\boldsymbol{Z}^{(t)} - \alpha_t \boldsymbol{R}^{(t)}\|_F^2} \mathcal{A}(\boldsymbol{Z}^{(t)} - \alpha_t \boldsymbol{R}^{(t)}) - \boldsymbol{y}, \mathcal{A}(\boldsymbol{R}^{(t)}) - \boldsymbol{y}\right\rangle\right|. \tag{3.33c}$$

The minimization of (3.33d) over α_t can be done by an exhaustive search over a small neighborhood of zero.

The third approach is to set α_t as the asymptotic limit given in (3.19a). We next provide numerical simulations to show the impact of the above three different choices of α_t on the performance of TARM.

3.4.3 *Numerical Results*

In this subsection, we compare the performance of TARM algorithms with different choices of μ_t and α_t. We also compare TARM with the existing matrix completion algorithms, including RCG [5], RGrad [5], NIHT [4], ALPS [17], LMAFit [20], and LRGeomCG [18]. The matrix form $\boldsymbol{A} \in \mathbb{R}^{m\times n}$ of the matrix completion operator $\mathcal{A}$ is chosen as a random selection matrix (with randomly selected rows from a permutation matrix). The low-rank matrix $\boldsymbol{X}_0 \in \mathbb{R}^{n_1\times n_2}$ is generated by the multiplication of two random Gaussian matrices of size $n_1 \times r$ and $r \times n_2$.

3.4.3.1 Non-Gaussianity of the Output Error of Module A

In Fig. 3.6, we plot the QQ plot of the input estimation errors of Module A of TARM for matrix completion. The QQ plot shows that the distribution of the estimation errors of Module A is non-Gaussian. Thus, Assumption 3.2 does not hold for matrix completion.

3.4.3.2 Comparisons of Different Choices of μ_t

We compare the TARM algorithms with μ_t in (3.29) and the subspace $\mathcal{S}$ given by (3.30), as shown in Fig. 3.7. We see that the performance of TARM is not sensitive to the three choices of $\mathcal{S}$ in (3.30). In the following, we always choose μ_t with $\mathcal{S}$ given by (3.30a).

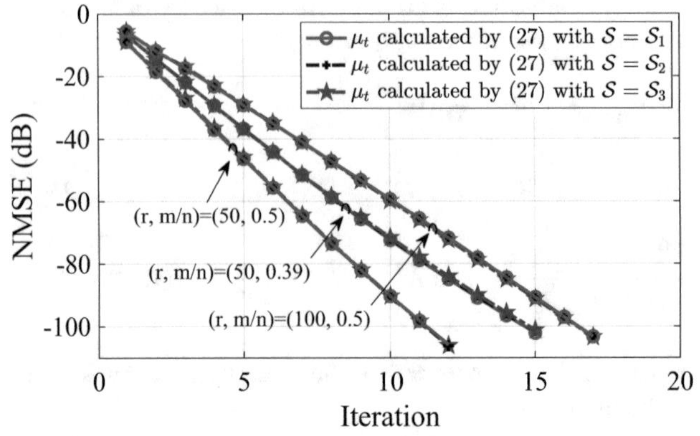

Fig. 3.7 Comparison of the TARM algorithms for matrix completion with different choices of μ_t. $n_1 = n_2 = 1000, \sigma^2 = 0$

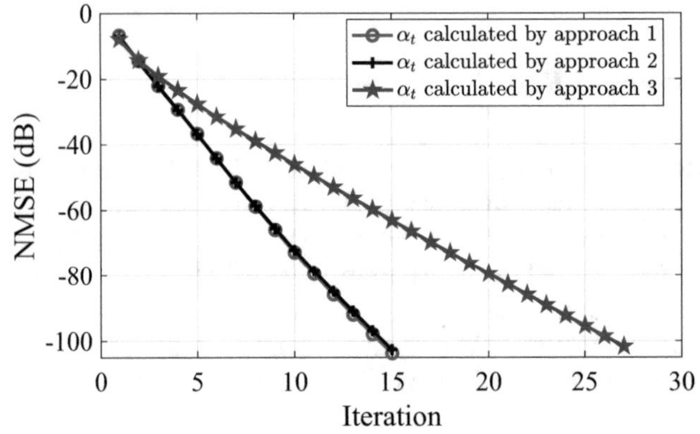

Fig. 3.8 Comparison of the TARM algorithms for matrix completion with different choices of α_t. $n_1 = n_2 = 1000, r = 50, \frac{m}{n_1 n_2} = 0.39, \sigma^2 = 0$

3.4.3.3 Comparisons of Different Choices of α_t

We compare the TARM algorithms with α_t given by the three different approaches in Appendix 2. As shown in Fig. 3.8, the first two approaches perform close to each other; the third approach performs considerably worse than the first two. Note that the first approach involves the computation of the divergence in (3.32), which is computationally demanding. Thus, we henceforth choose α_t based on the second approach in (3.33).

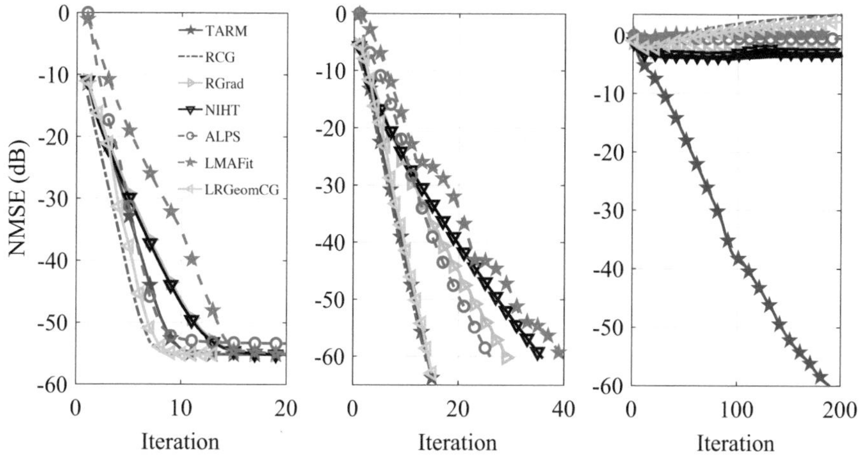

Fig. 3.9 Comparison of algorithms for matrix completion. Left: $n_1 = n_2 = 1000, r = 50, m/n = 0.39, \sigma^2 = 10^{-5}$. Middle: $n_1 = n_2 = 1000, r = 20, m/n = 0.12, \sigma^2 = 0$. Right: $n_1 = n_2 = 1000, r = 20, m/n = 0.045, \sigma = 0$

3.4.3.4 Performance Comparisons

We compare TARM with the existing algorithms for matrix completion in Fig. 3.9. We see that in the left and middle plots (with measurement rate $m/n = 0.39$ and 0.12), TARM performs close to LRGeomCG and RCG, while in the right plot (with $m/n = 0.045$), TARM significantly outperforms the other algorithms. More detailed performance comparisons are given in Sect. 3.5.

3.4.3.5 Empirical Phase Transition

Similar to the case of low-rank matrix recovery. We consider an algorithm to be successful in recovering the low-rank matrix $\boldsymbol{X}_0$ when the following conditions are satisfied: (1) the normalized mean square error $\frac{\|\boldsymbol{X}^{(t)} - \boldsymbol{X}_0\|_F^2}{\|\boldsymbol{X}_0\|_F^2} \leq 10^{-6}$; (2) the iteration number $t < 1000$. In Fig. 3.5, we plot the phase transition curves of the algorithms mentioned before. From Fig. 3.10, we see that the phase transition of TARM is the closest to the upper bound and considerably higher than the curves of NIHT and RGrad.

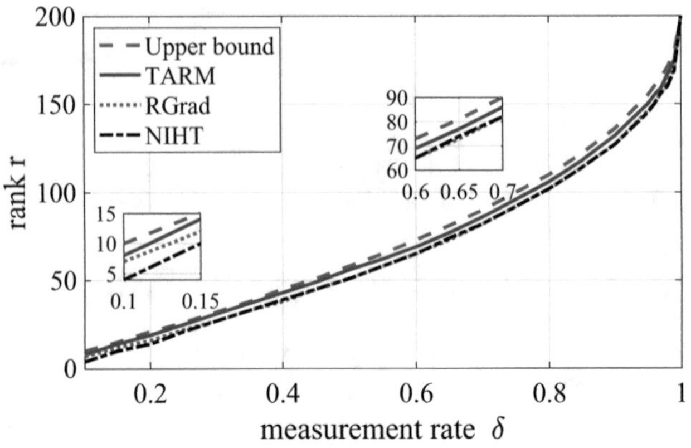

Fig. 3.10 The phase transition curves of various matrix completion algorithms. $n_1 = n_2 = 200$, $\sigma^2 = 0$. For each algorithm, the region below the phase transition curve corresponds to the successful recovery of $\boldsymbol{X}_0$

3.5 More Numerical Results

We now compare the performance of TARM with other existing algorithms including LMAFit [20], RCG [5], ALPS [17], LRGeomCG [18], BARM [21], and IRLS0 [22] for low-rank matrix recovery and matrix completion problems. In our comparisons, we set $n_1 = n_2$, $n = n_1 n_2$, and $\sigma = 0$. All experiments are conducted in Matlab on an Intel 2.3 GHz Core i5 Quad-core processor with 16 GB RAM on MacOS. The implementation of ALPS, BARM, and IRLS0 is from public sources and the implementation of LMAFit, RCG, RGrad, LRGeomCG, and NIHT is from our own codes (available for downloading at GitHub). For a fair comparison, all the codes are realized in a pure Matlab environment without using MEX files for acceleration. The comparison results are shown in Tables 3.2, 3.3, and 3.4. In these tables, "#iter" denotes the average iteration times of a successful recovery, "NS" denotes the number of successful recovery out of 10 trials for each settings, and "Time" denotes the average running time of successful recoveries.

3.5.1 Low-Rank Matrix Recovery

3.5.1.1 Algorithm Comparison

In Table 3.2 we compare the performance of algorithms for low-rank matrix recovery with different settings. The linear operator is chosen as the partial orthogonal operator in (3.28). LRGeomCG is not included for that it only applies to matrix completion. From Table 3.2, we see that TARM has the best performance (with the least running time and the highest success rate), and LMAFit does not work

Table 3.2 Comparisons of algorithms for low-rank matrix recovery

Parameters			TARM			LMAFit			RCG			ALPS			NIHT			RGrad		
n_1	m/n	r	#iter	NS	Time	#iter	NS	Time	#iter	NS	Time	#iter	NS	Time	#iter	NS	Time	#iter	NS	Time
1000	0.1	20	14	10	**4.18**	–	0	–	14	10	6.65	9	10	4.55	38.2	10	9.54	38	10	8.66
1000	0.07	20	23	10	**6.98**	–	0	–	23.7	10	11.12	30.6	10	15.75	96.8	10	24.48	95.9	10	21.57
1000	0.05	20	54.4	10	**15.89**	–	0	–	60.2	10	29.05	–	0	–	–	0	–	–	0	–
1000	0.045	20	96.2	10	**28.43**	–	0	–	119.4	10	55.69	–	0	–	–	0	–	–	0	–
1000	0.042	20	198	9	**61.90**	–	0	–	–	0	–	–	0	–	–	0	–	–	0	–
1000	0.1	30	24	10	**9.10**	–	0	–	25	10	12.14	34	10	23.29	105	10	32.40	105	10	25.36
1000	0.2	50	16	10	**10.62**	–	0	–	17	10	10.13	12	10	12.83	51	10	22.25	51	10	13.97

The least values of time are highlighted with a bold font

Table 3.3 Comparisons of TARM with LMAFit, RCG, and ALPS for matrix completions

Parameters			TARM			LMAFit			RCG			ALPS			NIHT			RGrad			LRGeomCG		
n_1	m/n	r	#iter	NS	Time	#iter	NS	Time	#iter	NS	Time	#iter	NS	Time	#iter	NS	Time	#iter	NS	Time	#iter	NS	Time
500	0.2	10	10.2	10	0.24	21.4	10	**0.076**	9.6	10	0.33	11.5	10	0.212	18.7	10	0.192	17.5	10	0.295	10	10	0.119
500	0.12	10	20.7	10	0.41	48	4	**0.13**	17.5	10	0.54	33.6	10	0.66	49.8	10	0.47	32.2	10	0.53	17.2	10	0.16
500	0.05	10	199.3	6	**9.05**	–	0	–	–	0	–	–	0	–	0	–	0	–	0	–	–	0	–
1000	0.12	10	10	10	0.97	21.7	10	**0.208**	9	10	1.99	13.4	10	0.93	17.2	10	0.69	15.1	10	1.55	9.4	10	0.46
1000	0.06	10	21.6	10	2.08	–	0	–	18.4	10	4.18	46.1	10	3.65	52.1	8	1.89	37	10	3.88	18.2	10	**0.78**
1000	0.026	10	195.5	8	**21.82**	–	0	–	–	0	–	–	0	–	0	–	0	–	0	–	–	0	–
1000	0.12	20	14.4	10	1.57	39.7	10	**0.41**	14	10	3.18	25.8	10	3.02	35.7	10	1.79	28.5	10	3.10	14.2	10	0.77
1000	0.06	20	48.1	10	5.37	–	0	–	48.9	10	11.55	–	0	–	295	2	12.89	185	9	20.08	48.1	10	**2.23**
1000	0.045	20	185	7	**30.48**	–	0	–	–	0	–	–	0	–	–	0	–	–	0	–	–	0	–
2000	0.12	40	15	10	11.78	33.4	10	**1.77**	13	10	23.6	20	10	17.10	31.7	10	11.26	29	10	23.37	13	10	4.78
2000	0.06	40	45.4	10	34.98	–	0	–	38.6	10	73.3	–	0	–	230	8	74.04	174.6	10	140.06	39	10	**12.31**
2000	0.044	40	145.5	10	**116.07**	–	0	–	–	0	–	–	0	–	0	–	0	–	0	–	–	0	–

The least values of time are highlighted with a bold font

Table 3.4 Comparisons of TARM with BARM and IRLS0 for matrix completion

Parameters			TARM			BRAM			IRLS0		
n_1	m/n	r	#iter	NS	Time	#iter	NS	Time	#iter	NS	Time
100	0.2	5	44	10	1.43	45	10	9.43	547	10	**0.58**
100	0.2	8	157	10	5.39	66	10	13.83	6102	10	**4.1373**
150	0.2	10	47	10	1.85	48	10	62.36	917	10	**1.08**
150	0.15	10	159	10	**5.93**	94	10	78.25	–	0	–
200	0.12	10	295.5	9	**12.01**	–	0	–	–	0	–

The least values of time are highlighted with a bold font

well in the considered settings. It is worth noting that TARM works well at low measurement rates when the other algorithms fail in recovery.

3.5.1.2 Impact of the Singular Value Distribution of the Low-Rank Matrix on TARM

The low-rank matrix generated by the product of two Gaussian matrices has a clear singular-value gap, i.e., the smallest singular σ_r is not close to zero. We now discuss the impact of the singular value distribution of the low-rank matrix on the performance of TARM.

To this end, we generate two matrices $\boldsymbol{G}_1 \in \mathbb{R}^{n_1 \times r}$ and $\boldsymbol{G}_2 \in \mathbb{R}^{r \times n_2}$ with the elements independently drawn from $\mathcal{N}(0, 1)$. Let the SVDs of $\boldsymbol{G}_1$ and $\boldsymbol{G}_2$ be $\boldsymbol{G}_1 = \boldsymbol{U}_1\boldsymbol{\Sigma}_1\boldsymbol{V}_1^T$, $\boldsymbol{G}_2 = \boldsymbol{U}_2\boldsymbol{\Sigma}_2\boldsymbol{V}_2^T$, where $\boldsymbol{U}_i \in \mathbb{R}^{n_i \times r}$ and $\boldsymbol{V}_i \in \mathbb{R}^{r \times n_i}$ are the left and right singular vector matrices, respectively, and $\boldsymbol{\Sigma}_i \in \mathbb{R}^{r \times r}$ is the singular value matrix for $i = 1, 2$. The low-rank matrix $\boldsymbol{X}_0$ is then generated as

$$\tilde{\boldsymbol{X}} = \boldsymbol{U}_1 \text{diag}(\exp(-k), \exp(-2k), \ldots, \exp(-rk))\boldsymbol{U}_2^T \tag{3.34}$$

where k controls the rate of decay. The matrix $\tilde{\boldsymbol{X}}$ is normalized to yield $\boldsymbol{X}_0 = \frac{\sqrt{n}\tilde{\boldsymbol{X}}}{\|\tilde{\boldsymbol{X}}\|_F}$. It is readily seen that the singular values of $\boldsymbol{X}_0$ do not have a clear gap away from zero, provided that k is sufficiently large.

In simulation, we consider the low-rank matrix recovery problem with the following three decay rates: $k = 0.1$, $k = 0.5$, and $k = 1$. We choose matrix dimensions $n_1 = n_2 = 1000$, rank $r = 20$, and measurement rate $\frac{m}{n} = 0.08$. The singular value distributions of low-rank matrix $\boldsymbol{X}_0$ in the three cases are given in the left figure of Fig. 3.11. We see that the singular values for $k = 0.1$ have a clear gap away from 0, while the singular values for $k = 0.5$ and 1 decay to 0 rapidly. In TARM, the target rank is set to the real rank value of $\boldsymbol{X}_0$. The NMSE of TARM against the iteration number is plotted on the right part of Fig. 3.11. We see that TARM converges the fastest for $k = 0.1$, and it works still well for $k = 0.5$ and 1 but with reduced convergence rates. Generally speaking, the convergence rate of TARM becomes slower as the increase of k.

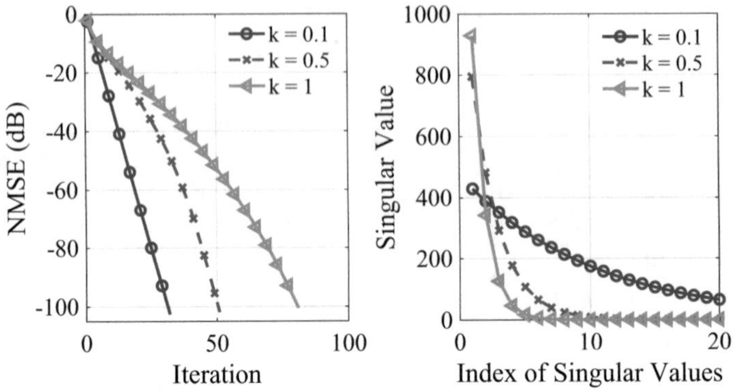

Fig. 3.11 The performance of the TARM algorithm when the low-rank matrices to be recovered do not have a clear singular-value gap away from zero

3.5.2 *Matrix Completion*

In Table 3.3, we compare the performance of algorithms for matrix completion in various settings. The linear operator is chosen as the random selection operator. We see from Table 3.3 that LMAFit runs the fastest when the measurement rate m/n is relatively high (say, $m/n = 0.2$ with $n_1 = n_2 = 500$); LRGeomCG performs the best when the measurement rate is medium (say, $m/n = 0.06$ with $n_1 = n_2 = 1000$); and TARM works the best when the measurement rate is relatively low (say, $m/n = 0.026$ with $n_1 = n_2 = 1000$).

Since both BARM and IRLS0 involve matrix inversions, these algorithms cannot handle the ARM problem with a relatively large size. Therefore, we set a relatively small size in our comparisons. The comparison results of TARM with BARM and IRLS0 on matrix completion are presented in Table 3.4. From the table, we see that TARM generally requires a shorter recovery time and has a higher recovery success rate than the other two counterparts.

3.6 Summary

In this chapter, we proposed a low-complexity iterative algorithm termed TARM for solving the stable ARM problem. The proposed algorithm can be applied to both low-rank matrix recovery and matrix completion. For low-rank matrix recovery, the performance of TARM can be accurately characterized by the state evolution technique when ROIL operators are involved. For matrix completion, we showed that, although state evolution is not accurate, the parameters of TARM can be carefully tuned to achieve good performance. Numerical results demonstrate that TARM has competitive performance compared to the existing algorithms for low-rank matrix recovery problems with ROIL operators and random selection operators.

Appendix 1: Proof of Lemma 3.1

We first determine μ_t. We have

$$\begin{aligned}
&\left\langle \boldsymbol{R}^{(t)} - \boldsymbol{X}_0, \boldsymbol{X}^{(t-1)} - \boldsymbol{X}_0 \right\rangle \\
&= \left\langle \boldsymbol{X}^{(t-1)} + \mu_t \mathcal{A}^T(\boldsymbol{y} - \mathcal{A}(\boldsymbol{X}^{(t-1)})) - \boldsymbol{X}_0, \boldsymbol{X}^{(t-1)} - \boldsymbol{X}_0 \right\rangle && \text{(3.35a)} \\
&= \left\langle \boldsymbol{X}^{(t-1)} + \mu_t \mathcal{A}^T(\mathcal{A}(\boldsymbol{X}_0) + \boldsymbol{n} - \mathcal{A}(\boldsymbol{X}^{(t-1)})) - \boldsymbol{X}_0, \boldsymbol{X}^{(t-1)} - \boldsymbol{X}_0 \right\rangle && \text{(3.35b)} \\
&= \left\langle \boldsymbol{X}^{(t-1)} - \boldsymbol{X}_0, \boldsymbol{X}^{(t-1)} - \boldsymbol{X}_0 \right\rangle + \mu_t \left\langle \boldsymbol{n}, \mathcal{A}(\boldsymbol{X}^{(t-1)} - \boldsymbol{X}_0) \right\rangle \\
&\quad - \mu_t \left\langle \mathcal{A}(\boldsymbol{X}^{(t-1)} - \boldsymbol{X}_0), \mathcal{A}(\boldsymbol{X}^{(t-1)} - \boldsymbol{X}_0) \right\rangle && \text{(3.35c)}
\end{aligned}$$

where step (3.35a) follows by substituting $\boldsymbol{R}^{(t)}$ in Line 3 of Algorithm 3, and step (3.35c) follows by noting

$$\langle \mathcal{A}(\boldsymbol{B}), \boldsymbol{c} \rangle = \left\langle \boldsymbol{B}, \mathcal{A}^T(\boldsymbol{c}) \right\rangle \tag{3.36}$$

for any matrix $\boldsymbol{B}$ and vector $\boldsymbol{c}$ of appropriate sizes. Together with Condition 1, we obtain (3.10a).

We next determine α_t and c_t. First note

$$\begin{aligned}
\|\boldsymbol{X}^{(t)} - \boldsymbol{R}^{(t)}\|_F^2 &= \|\boldsymbol{X}^{(t)} - \boldsymbol{X}_0\|_F^2 + \|\boldsymbol{X}_0 - \boldsymbol{R}^{(t)}\|_F^2 \\
&\quad + 2\left\langle \boldsymbol{X}^{(t)} - \boldsymbol{X}_0, \boldsymbol{X}_0 - \boldsymbol{R}^{(t)} \right\rangle && \text{(3.37a)} \\
&= \|\boldsymbol{X}^{(t)} - \boldsymbol{X}_0\|_F^2 + \|\boldsymbol{X}_0 - \boldsymbol{R}^{(t)}\|_F^2 && \text{(3.37b)}
\end{aligned}$$

where (3.37b) is from Condition 2 in (3.8). Recall that in the t-th iteration $\boldsymbol{R}^{(t)}$ is a function of μ_t but not of α_t and c_t. Thus, minimizing $\|\boldsymbol{X}^{(t)} - \boldsymbol{X}_0\|_F^2$ over α_t and c_t is equivalent to minimizing $\|\boldsymbol{X}^{(t)} - \boldsymbol{R}^{(t)}\|_F^2$ over α_t and c_t. For any given α_t, the optimal c_t to minimize $\|\boldsymbol{X}^{(t)} - \boldsymbol{R}^{(t)}\|_F^2 = \|c_t(\boldsymbol{Z}^{(t)} - \alpha_t \boldsymbol{R}^{(t)}) - \boldsymbol{R}^{(t)}\|_F^2$ is given by

$$c_t = \frac{\left\langle \boldsymbol{Z}^{(t)} - \alpha_t \boldsymbol{R}^{(t)}, \boldsymbol{R}^{(t)} \right\rangle}{\|\boldsymbol{Z}^{(t)} - \alpha_t \boldsymbol{R}^{(t)}\|_F^2}. \tag{3.38}$$

Then,

$$\begin{aligned}&\left\langle \boldsymbol{X}^{(t)} - \boldsymbol{X}_0, \boldsymbol{R}^{(t)} - \boldsymbol{X}_0 \right\rangle \\ &= \left\langle c_t(\boldsymbol{Z}^{(t)} - \alpha_t \boldsymbol{R}^{(t)}) - \boldsymbol{X}_0, \boldsymbol{R}^{(t)} - \boldsymbol{X}_0 \right\rangle &\text{(3.39a)} \\ &= \left\langle \frac{\langle \boldsymbol{Z}^{(t)} - \alpha_t \boldsymbol{R}^{(t)}, \boldsymbol{R}^{(t)} \rangle}{\|\boldsymbol{Z}^{(t)} - \alpha_t \boldsymbol{R}^{(t)}\|_F^2} (\boldsymbol{Z}^{(t)} - \alpha_t \boldsymbol{R}^{(t)}) - \boldsymbol{X}_0, \boldsymbol{R}^{(t)} - \boldsymbol{X}_0 \right\rangle &\text{(3.39b)}\end{aligned}$$

where (3.39a) follows by substituting $\boldsymbol{X}^{(t)}$ in Line 5 of Algorithm 3, and (3.39b) by substituting c_t in (3.38). Combining (3.39) and Condition 2, we see that α_t is the solution of the following quadratic equation:

$$a_t \alpha_t^2 + b_t \alpha_t + d_t = 0 \tag{3.40}$$

where a_t, b_t, and d_t are defined in (3.11). Therefore, α_t is given by (3.10b). With the above choice of c_t, we have

$$\begin{aligned}&\left\langle \boldsymbol{X}^{(t)} - \boldsymbol{R}^{(t)}, \boldsymbol{X}^{(t)} \right\rangle \\ &= \left\langle c_t(\boldsymbol{Z}^{(t)} - \alpha_t \boldsymbol{R}^{(t)}) - \boldsymbol{R}^{(t)}, c_t(\boldsymbol{Z}^{(t)} - \alpha_t \boldsymbol{R}^{(t)}) \right\rangle = 0. &\text{(3.41)}\end{aligned}$$

This orthogonality is useful in analyzing the performance of Module B.

Appendix 2: Convergence Analysis of TARM Based on RIP

Without loss of generality, we assume $n_1 \leq n_2$ in this appendix. Following the convention in [4], we focus our discussion on the noiseless case, i.e., $\boldsymbol{n} = \boldsymbol{0}$.

Definition 3.2 (Restricted Isometry Property) Given a linear operator $\mathcal{A} : \mathbb{R}^{n_1 \times n_2} \to \mathbb{R}^m$, a minimum constant called the rank restricted isometry constant (RIC) $\delta_r(\mathcal{A}) \in (0, 1)$ exists such that

$$(1 - \delta_r(\mathcal{A}))\|\boldsymbol{X}\|_F^2 \leq \|\gamma \mathcal{A}(\boldsymbol{X})\|_2^2 \leq (1 + \delta_r(\mathcal{A}))\|\boldsymbol{X}\|_F^2 \tag{3.42}$$

for all $\boldsymbol{X} \in \mathbb{R}^{n_1 \times n_2}$ with $\text{rank}(\boldsymbol{X}) \leq r$, where $\gamma > 0$ is a constant scaling factor.

We now introduce two useful lemmas.

Lemma 3.3 *Assume that α_{t+1} and c_{t+1} satisfy Conditions 2 and 3. Then,*

$$\|\boldsymbol{X}^{(t)} - \boldsymbol{R}^{(t)}\|_F^2 = \frac{\|\boldsymbol{R}^{(t)} - \boldsymbol{Z}^{(t)}\|_F^2}{\frac{\|\boldsymbol{R}^{(t)} - \boldsymbol{Z}^{(t)}\|_F^2}{\|\boldsymbol{Z}^{(t)}\|_F^2}\alpha_t^2 + (1-\alpha_t)^2}. \tag{3.43}$$

Lemma 3.4 *Let $\boldsymbol{Z}^{(t)}$ be the best rank-r approximation of $\boldsymbol{R}^{(t)}$. Then,*

$$\|\boldsymbol{R}^{(t)} - \boldsymbol{Z}^{(t)}\|_F^2 \le \|\boldsymbol{X}_0 - \boldsymbol{R}^{(t)}\|_F^2. \tag{3.44}$$

The proof of Lemma 3.3 is given in Appendix 5. Lemma 3.4 is straightforward from the definition of the best rank-r approximation [8, p. 211–218].

Theorem 3.3 *Assume that μ_t, α_t, c_t satisfy Conditions 1–3, and the linear operator $\mathcal{A}$ satisfies the RIP with rank n_1 and RIC δ_{n_1}. Then,*

$$\|\boldsymbol{X}^{(t)} - \boldsymbol{X}_0\|_F^2 \le \left(\frac{1}{(1-\alpha_t)^2} - 1\right)\left(\frac{1+\delta_{n_1}}{1-\delta_{n_1}} - 1\right)^2 \|\boldsymbol{X}^{(t-1)} - \boldsymbol{X}_0\|_F^2 \tag{3.45}$$

TARM guarantees to converge when RIC satisfies $\alpha_t \neq 1, \forall t$, and

$$\delta_{n_1} < \frac{1}{1 + 2\sqrt{\frac{1}{\xi}\left(\frac{1}{(1-\alpha_{max})^2} - 1\right)}} \tag{3.46}$$

where the constant ξ satisfies $0 < \xi < 1$, and $\alpha_{max} = \sup\{\alpha_t\}$.

Proof Since $\boldsymbol{Z}^{(t)}$ is the best rank-r approximation of $\boldsymbol{R}^{(t)}$, we have $\|\boldsymbol{R}^{(t)}\|_F^2 \ge \|\boldsymbol{Z}^{(t)}\|_F^2$. Then, from Lemma 3.3, we obtain

$$\|\boldsymbol{X}^{(t)} - \boldsymbol{R}^{(t)}\|_F^2 \le \frac{\|\boldsymbol{R}^{(t)} - \boldsymbol{Z}^{(t)}\|_F^2}{(1-\alpha_t)^2}. \tag{3.47}$$

Then, we have

$$\|\boldsymbol{X}^{(t)} - \boldsymbol{R}^{(t)}\|_F^2 = \|\boldsymbol{X}^{(t)} - \boldsymbol{X}_0 + \boldsymbol{X}_0 - \boldsymbol{R}^{(t)}\|_F^2 \tag{3.48a}$$

$$= \|\boldsymbol{X}^{(t)} - \boldsymbol{X}_0\|_F^2 + \|\boldsymbol{X}_0 - \boldsymbol{R}^{(t)}\|_F^2 + 2\left\langle \boldsymbol{X}^{(t)} - \boldsymbol{X}_0, \boldsymbol{X}_0 - \boldsymbol{R}^{(t)}\right\rangle \tag{3.48b}$$

$$= \|\boldsymbol{X}^{(t)} - \boldsymbol{X}_0\|_F^2 + \|\boldsymbol{X}_0 - \boldsymbol{R}^{(t)}\|_F^2 \tag{3.48c}$$

where (3.48c) follows from $\left\langle \boldsymbol{X}^{(t)} - \boldsymbol{X}_0, \boldsymbol{X}_0 - \boldsymbol{R}^{(t)}\right\rangle = 0$ in Condition 2. Combining (3.4), (3.47), and (3.48), we obtain

$$\|\boldsymbol{X}^{(t)} - \boldsymbol{X}_0\|_F^2 \leq \left(\frac{1}{(1-\alpha_t)^2} - 1\right) \|\boldsymbol{R}^{(t)} - \boldsymbol{X}_0\|_F^2 \tag{3.49a}$$

$$= \left(\frac{1}{(1-\alpha_t)^2} - 1\right) \|\boldsymbol{X}^{(t-1)} + \mu_t \mathcal{A}^*(\boldsymbol{y} - \mathcal{A}(\boldsymbol{X}^{(t-1)})) - \boldsymbol{X}_0\|_F^2 \tag{3.49b}$$

$$= \left(\frac{1}{(1-\alpha_t)^2} - 1\right) \|(\mathcal{I} - \mu_t \mathcal{A}^* \mathcal{A})(\boldsymbol{X}^{(t-1)} - \boldsymbol{X}_0)\|_F^2. \tag{3.49c}$$

Since $\mathcal{A}$ has RIP with rank n_1 and RIC δ_{n_1}, we obtain the following inequality from [17]:

$$\|(\mathcal{I} - \mu_t \mathcal{A}^* \mathcal{A})(\boldsymbol{X}^{(t-1)} - \boldsymbol{X}_0)\|_F^2$$
$$\leq \max\left((\mu_t(1+\delta_{n_1}) - 1)^2, (\mu_t(1-\delta_{n_1}) - 1)^2\right) \|\boldsymbol{X}^{(t-1)} - \boldsymbol{X}_0\|_F^2. \tag{3.50}$$

Recall that $\mu_t = \frac{\|\boldsymbol{X}^{(t-1)} - \boldsymbol{X}_0\|_F^2}{\|\mathcal{A}(\boldsymbol{X}^{(t-1)} - \boldsymbol{X}_0)\|_2^2}$ obtained by letting $\boldsymbol{n} = \boldsymbol{0}$ in (3.10a). From RIP, we have

$$\frac{1}{1+\delta_{n_1}} \leq \mu_t = \frac{\|\boldsymbol{X}^{(t-1)} - \boldsymbol{X}_0\|_F^2}{\|\mathcal{A}(\boldsymbol{X}^{(t-1)} - \boldsymbol{X}_0)\|_2^2} \leq \frac{1}{1-\delta_{n_1}}. \tag{3.51}$$

Then, combining (3.50) and (3.51), we have

$$\|(\mathcal{I} - \mu_t \mathcal{A}^* \mathcal{A})(\boldsymbol{X}^{(t-1)} - \boldsymbol{X}_0)\|_F^2 \leq \left(\frac{1+\delta_{n_1}}{1-\delta_{n_1}} - 1\right)^2 \|\boldsymbol{X}^{(t-1)} - \boldsymbol{X}_0\|_F^2. \tag{3.52}$$

Combining (3.52) and (3.49), we arrive at (3.45).

When δ_{n_1} satisfies (3.46), we have

$$\|\boldsymbol{X}^{(t)} - \boldsymbol{X}_0\|_F^2 < \xi \|\boldsymbol{X}^{(t-1)} - \boldsymbol{X}_0\|_F^2 \tag{3.53}$$

at each iteration t. Then, TARM converges exponentially to $\boldsymbol{X}_0$. □

We now compare the convergence rate of TARM with those of SVP and NIHT. Compared with [4, Eq. 2.11–2.14], (3.45) contains an extra term $\frac{1}{(1-\alpha_t)^2} - 1$. From numerical experiments, α_t is usually close to zero, implying that TARM converges faster than SVP and NIHT.

Appendix 3: Proof of Theorem 3.1

For a partial orthogonal ROIL operator $\mathcal{A}$, the following properties hold:

$$\mathcal{A}(\mathcal{A}^T(\boldsymbol{a})) = \boldsymbol{a} \tag{3.54a}$$

$$\left\langle \mathcal{A}^T(\boldsymbol{a}), \mathcal{A}^T(\boldsymbol{b}) \right\rangle = \langle \boldsymbol{a}, \boldsymbol{b} \rangle . \tag{3.54b}$$

Then as $m, n \to \infty$ with $\frac{m}{n} \to \delta$, we have

$$\begin{aligned}
&\left\| \boldsymbol{R}^{(t)} - \boldsymbol{X}_0 \right\|_F^2 \\
&= \left\| \boldsymbol{X}^{(t-1)} - \boldsymbol{X}_0 - \frac{1}{\delta}\mathcal{A}^T \mathcal{A}(\boldsymbol{X}^{(t-1)} - \boldsymbol{X}_0) + \frac{1}{\delta}\mathcal{A}^T(\boldsymbol{n}) \right\|_F^2 && (3.55a) \\
&= \|\boldsymbol{X}^{(t-1)} - \boldsymbol{X}_0\|_F^2 + \frac{1}{\delta^2}\|\mathcal{A}(\boldsymbol{X}^{(t-1)} - \boldsymbol{X}_0)\|_F^2 \\
&\quad - \frac{2}{\delta}\|\mathcal{A}(\boldsymbol{X}^{(t-1)} - \boldsymbol{X}_0)\|_F^2 + \frac{1}{\delta^2}\|\boldsymbol{n}\|_2^2 && (3.55b) \\
&= \|\boldsymbol{X}^{(t-1)} - \boldsymbol{X}_0\|_F^2 + \frac{1}{\delta}\|\boldsymbol{X}^{(t-1)} - \boldsymbol{X}_0\|_F^2 \\
&\quad - 2\|\boldsymbol{X}^{(t-1)} - \boldsymbol{X}_0\|_F^2 + \frac{1}{\delta^2}\|\boldsymbol{n}\|_2^2 && (3.55c) \\
&= \left(\frac{1}{\delta} - 1\right)\|\boldsymbol{X}^{(t-1)} - \boldsymbol{X}_0\|_F^2 + 1/\delta n \sigma^2 && (3.55d)
\end{aligned}$$

where (3.55a) is obtained by substituting $\boldsymbol{R}^{(t)} = \boldsymbol{X}^{(t-1)} + \mu_t \mathcal{A}^T(\boldsymbol{y} - \mathcal{A}(\boldsymbol{X}^{(t-1)}))$ and $\boldsymbol{y} = \mathcal{A}(\boldsymbol{X}_0) + \boldsymbol{n}$ with $\mu_t = \delta^{-1}$, (3.55b) is obtained by noting that $\boldsymbol{n}$ is independent of $\mathcal{A}(\boldsymbol{X}^{(t)} - \boldsymbol{X}_0)$ (ensured by Assumption 3.1) and (3.54b), and (3.55c) follows from $\frac{\|\mathcal{A}(\boldsymbol{X}^{(t-1)} - \boldsymbol{X}_0)\|_2^2}{\|\boldsymbol{X}^{(t-1)} - \boldsymbol{X}_0\|_F^2} \to \delta$ (see (3.13)). When $\frac{1}{n}\|\boldsymbol{X}^{(t-1)} - \boldsymbol{X}_0\|_F^2 \to \tau$, we have

$$\frac{1}{n}\|\boldsymbol{R}^{(t)} - \boldsymbol{X}_0\|_F^2 \to \left(\frac{1}{\delta} - 1\right)\tau + 1/\delta\sigma^2. \tag{3.56}$$

We now consider the case of Gaussian ROIL operators. As $m, n \to \infty$ with $\frac{m}{n} \to \delta$, we have

$$\begin{aligned}
&\left\| \boldsymbol{R}^{(t)} - \boldsymbol{X}_0 \right\|_F^2 \\
&= \left\| \boldsymbol{X}^{(t-1)} - \boldsymbol{X}_0 - \frac{1}{\delta}\mathcal{A}^T \mathcal{A}(\boldsymbol{X}^{(t-1)} - \boldsymbol{X}_0) + \frac{1}{\delta}\mathcal{A}^T(\boldsymbol{n}) \right\|_F^2 && (3.57a) \\
&= \|\boldsymbol{X}^{(t-1)} - \boldsymbol{X}_0\|_F^2 + \frac{1}{\delta^2}\|\mathcal{A}^T\mathcal{A}(\boldsymbol{X}^{(t-1)} - \boldsymbol{X}_0)\|_F^2 \\
&\quad - \frac{2}{\delta}\|\mathcal{A}(\boldsymbol{X}^{(t-1)} - \boldsymbol{X}_0)\|_F^2 + \frac{1}{\delta^2}\|\boldsymbol{n}\|_2^2 && (3.57b)
\end{aligned}$$

$$= \|\boldsymbol{X}^{(t-1)} - \boldsymbol{X}_0\|_F^2 + \frac{1}{\delta^2}\|\boldsymbol{A}^T\boldsymbol{A}\text{vec}(\boldsymbol{X}^{(t-1)} - \boldsymbol{X}_0)\|_F^2$$

$$- \frac{2}{\delta}\|\mathcal{A}(\boldsymbol{X}^{(t-1)} - \boldsymbol{X}_0)\|_F^2 + \frac{1}{\delta^2}\|\boldsymbol{n}\|_2^2 \tag{3.57c}$$

$$= \|\boldsymbol{X}^{(t-1)} - \boldsymbol{X}_0\|_F^2 + \frac{1}{\delta^2}\frac{\|\boldsymbol{A}^T\boldsymbol{A}\|_F^2}{mn}\|\text{vec}(\boldsymbol{X}^{(t-1)} - \boldsymbol{X}_0)\|_2^2$$

$$- \frac{2}{\delta}\|\mathcal{A}(\boldsymbol{X}^{(t-1)} - \boldsymbol{X}_0)\|_F^2 + \frac{1}{\delta^2}\|\boldsymbol{n}\|_2^2 \tag{3.57d}$$

$$= \|\boldsymbol{X}^{(t-1)} - \boldsymbol{X}_0\|_F^2 + \frac{1}{\delta^2}\frac{\text{Tr}((\boldsymbol{A}^T\boldsymbol{A})^2)}{mn}\|\boldsymbol{X}^{(t-1)} - \boldsymbol{X}_0\|_F^2$$

$$- \frac{2}{\delta}\|\mathcal{A}(\boldsymbol{X}^{(t-1)} - \boldsymbol{X}_0)\|_F^2 + \frac{1}{\delta^2}\|\boldsymbol{n}\|_2^2 \tag{3.57e}$$

$$= \|\boldsymbol{X}^{(t-1)} - \boldsymbol{X}_0\|_F^2 + \left(1 + \frac{1}{\delta}\right)\|\boldsymbol{X}^{(t-1)} - \boldsymbol{X}_0\|_F^2$$

$$- \frac{2}{\delta}\|\mathcal{A}(\boldsymbol{X}^{(t-1)} - \boldsymbol{X}_0)\|_F^2 + \frac{1}{\delta^2}\|\boldsymbol{n}\|_2^2 \tag{3.57f}$$

$$= \|\boldsymbol{X}^{(t-1)} - \boldsymbol{X}_0\|_F^2 + \left(1 + \frac{1}{\delta}\right)\|\boldsymbol{X}^{(t-1)} - \boldsymbol{X}_0\|_F^2$$

$$- 2\|\boldsymbol{X}^{(t-1)} - \boldsymbol{X}_0\|_F^2 + \frac{1}{\delta^2}\|\boldsymbol{n}\|_2^2 \tag{3.57g}$$

$$= \frac{1}{\delta}\|\boldsymbol{X}^{(t-1)} - \boldsymbol{X}_0\|_F^2 + 1/\delta n\sigma^2 \tag{3.57h}$$

where (3.57a) is obtained by substituting $\boldsymbol{R}^{(t)} = \boldsymbol{X}^{(t-1)} + \mu_t\mathcal{A}^T(\boldsymbol{y} - \mathcal{A}(\boldsymbol{X}^{(t-1)}))$ and $\boldsymbol{y} = \mathcal{A}(\boldsymbol{X}_0) + \boldsymbol{n}$ with $\mu_t = \delta^{-1}$, (3.57b) is obtained by noting that $\boldsymbol{n}$ is independent of $\mathcal{A}(\boldsymbol{X}^{(t)} - \boldsymbol{X}_0)$ (from Assumption 3.1), (3.57c) follows by utilizing the matrix form $\boldsymbol{A}$ of $\mathcal{A}$, (3.57d) follows from the fact that $\boldsymbol{V}_A$ (i.e. the right singular vector matrix of $\boldsymbol{A}$) is a Haar distributed orthogonal matrix independent of $\boldsymbol{X}^{(t-1)} - \boldsymbol{X}_0$, (3.57e) follows from $\text{Tr}((\boldsymbol{A}^T\boldsymbol{A})^2) = \|\boldsymbol{A}^T\boldsymbol{A}\|_F^2$, (3.57f) is obtained by noting that $\frac{1}{mn}\text{Tr}((\boldsymbol{A}^T\boldsymbol{A})^2) \to \delta + \delta^2$ since $\boldsymbol{A}^T\boldsymbol{A}$ is a Wishart matrix with variance $\frac{1}{n}$ [11, p. 26], and (3.57g) follows by noting $\frac{\|\mathcal{A}(\boldsymbol{X}^{(t)} - \boldsymbol{X}_0)\|_2^2}{\|\boldsymbol{X}^{(t)} - \boldsymbol{X}_0\|_F^2} \to \delta$. When $\frac{1}{n}\|\boldsymbol{X}^{(t)} - \boldsymbol{X}_0\|_F^2 \to \tau$, we have

$$\frac{1}{n}\|\boldsymbol{R}^{(t)} - \boldsymbol{X}_0\|_F^2 \to \frac{1}{\delta}\tau + 1/\delta\sigma^2. \tag{3.58}$$

Appendix 4: Proof of Lemma 3.2

We first introduce two useful facts.

Fact 1: When $n_2 \to \infty$ with fixed $n_1/n_2 = \rho$, the i-th singular value σ_i of the Gaussian noise corrupted matrix $\boldsymbol{R}^{(t)}$ is given by [10, Eq. 9]

$$\frac{1}{\sqrt{n_2}}\sigma_i \xrightarrow{\text{a.s.}} \begin{cases} \sqrt{\frac{(v_t+\theta_i^2)(\rho v_t+\theta_i^2)}{\theta_i^2}} & \text{if } i \le r \text{ and } \theta_i > \rho^{\frac{1}{4}} \\ \sqrt{v_t}(1+\sqrt{\rho}) & \text{otherwise} \end{cases} \tag{3.59}$$

where v_t is the variance of the Gaussian noise, and θ_i is the i-th largest singular value of $\frac{1}{\sqrt{n_2}}\boldsymbol{X}_0$.

Fact 2: From [23, Eq. 9], the divergence of a spectral function $h(\boldsymbol{R})$ is given by

$$\begin{aligned} \operatorname{div}(h(\boldsymbol{R})) = |n_1 - n_2| \sum_{i=1}^{\min(n_1,n_2)} \frac{h_i(\sigma_i)}{\sigma_i} + \sum_{i=1}^{\min(n_1,n_2)} h_i'(\sigma_i) \\ + 2 \sum_{i \neq j, i,j=1}^{\min(n_1,n_2)} \frac{\sigma_i h_i(\sigma_i)}{\sigma_i^2 - \sigma_j^2}. \end{aligned} \tag{3.60}$$

The best rank-r approximation denoiser $\mathcal{D}(\boldsymbol{R})$ is a spectral function with

$$\begin{cases} h_i(\sigma_i) = \sigma_i & i \le r; \\ h_i(\sigma_i) = 0 & i > r. \end{cases} \tag{3.61}$$

Combining (3.60) and (3.61), the divergence of $\mathcal{D}(\boldsymbol{R}^{(t)})$ is given by

$$\operatorname{div}(\mathcal{D}(\boldsymbol{R}^{(t)})) = |n_1 - n_2| r + r^2 + 2 \sum_{i=1}^{r} \sum_{j=r+1}^{\min(n_1,n_2)} \frac{\sigma_i^2}{\sigma_i^2 - \sigma_j^2}. \tag{3.62}$$

Further, we have

$$\begin{aligned} &\sum_{i=1}^{r} \sum_{j=r+1}^{\min(n_1,n_2)} \frac{\sigma_i^2}{\sigma_i^2 - \sigma_j^2} \\ &\xrightarrow{a.s.} (\min(n_1, n_2) - r) \sum_{i=1}^{r} \frac{\sigma_i^2}{\sigma_i^2 - (\sqrt{n_2 v_t}(1+\sqrt{\rho}))^2} \end{aligned} \tag{3.63a}$$

$$= (\min(n_1, n_2) - r) \sum_{i=1}^{r} \frac{n_2 \frac{(v_t+\theta_i^2)(\rho v_t+\theta_i^2)}{\theta_i^2}}{\frac{n_2(v_t+\theta_i^2)(\rho v_t+\theta_i^2)}{\theta_i^2} - n_2 v_t (1+\sqrt{\rho})^2} \tag{3.63b}$$

$$= (\min(n_1, n_2) - r) \sum_{i=1}^{r} \frac{(v_t+\theta_i^2)(\rho v_t+\theta_i^2)}{(\sqrt{\rho} v_t - \theta_i^2)^2} \tag{3.63c}$$

$$\overset{a.s.}{\to} (\min(n_1, n_2) - r) r \int_0^\infty \frac{(v_t+\theta^2)(\rho v_t+\theta^2)}{(\sqrt{\rho} v_t - \theta^2)^2} p(\theta) d\theta \tag{3.63d}$$

$$= (\min(n_1, n_2) - r) r \Delta_1(v_t) \tag{3.63e}$$

where both (3.63a) and (3.63b) are from (3.59), and (3.63e) follows by the definition of $\Delta_1(v_t)$. Combining (3.62) and (3.63), we obtain the asymptotic divergence of $\mathcal{D}(\boldsymbol{R})$ given by

$$\text{div}(\mathcal{D}(\boldsymbol{R})) \overset{a.s.}{\to} |n_1 - n_2| r + r^2 + 2(\min(n_1, n_2) - r) r \Delta_1(v_t) \tag{3.64}$$

and

$$\alpha_t = \frac{1}{n} \text{div}(f(\boldsymbol{R}^{(t)})) \tag{3.65a}$$

$$\overset{a.s.}{\to} \left|1 - \frac{1}{\rho}\right| \lambda + \frac{1}{\rho}\lambda^2 + 2\left(\min\left(1, \frac{1}{\rho}\right) - \frac{\lambda}{\rho}\right) \lambda \Delta_1(v_t) \tag{3.65b}$$

$$= \alpha(v_t) \tag{3.65c}$$

with $\lambda = r/n_2$.

Recall that $\boldsymbol{Z}^{(t)}$ is the best rank-r approximation of $\boldsymbol{R}^{(t)}$ satisfying

$$\|\boldsymbol{Z}^{(t)}\|_F^2 = \sum_{i=1}^{r} \sigma_i^2 \tag{3.66a}$$

$$\|\boldsymbol{R}^{(t)}\|_F^2 - \|\boldsymbol{Z}^{(t)}\|_F^2 = \sum_{i=r+1}^{n_1} \sigma_i^2. \tag{3.66b}$$

Then, when $m, n \to \infty$ with $\frac{m}{n} \to \delta$, we have

$$\|\boldsymbol{Z}^{(t)}\|_F^2 = \sum_{i=1}^{r} \sigma_i^2 \tag{3.67a}$$

$$\xrightarrow{\text{a.s.}} n_2 \sum_{i=1}^{r} \frac{(v+\theta_i^2)(\rho v+\theta_i^2)}{\theta_i^2} \tag{3.67b}$$

$$= n + \lambda \left(1 + \frac{1}{\rho}\right) nv + \lambda n v^2 \frac{1}{r} \sum_{i=1}^{r} \frac{1}{\theta_i^2} \tag{3.67c}$$

and

$$\begin{aligned} &\|\boldsymbol{R}^{(t)}\|_F^2 - \|\boldsymbol{Z}^{(t)}\|_F^2 \\ &= \|\boldsymbol{X}_0\|_F^2 + nv_t - \|\boldsymbol{Z}^{(t)}\|_F^2 &(3.68a)\\ &\xrightarrow{\text{a.s.}} nv_t - \lambda(1 + \frac{1}{\rho})nv_t - \lambda n v_t^2 \frac{1}{r} \sum_{i=1}^{r} \frac{1}{\theta_i^2} &(3.68b) \end{aligned}$$

where (3.67b) is from (3.59), (3.68a) is from Assumption 3.2, and (3.68b) is from (3.67). Then,

$$\begin{aligned} c_t &= \frac{\langle \boldsymbol{Z}^{(t)} - \alpha_t \boldsymbol{R}^{(t)}, \boldsymbol{R}^{(t)} \rangle}{\|\boldsymbol{Z}^{(t)} - \alpha_t \boldsymbol{R}^{(t)}\|_F^2} &(3.69a)\\ &= \frac{\langle \boldsymbol{Z}^{(t)}, \boldsymbol{R}^{(t)} \rangle - \alpha_t \|\boldsymbol{R}^{(t)}\|_F^2}{\|\boldsymbol{Z}^{(t)}\|_F^2 - 2\alpha_t \langle \boldsymbol{Z}^{(t)}, \boldsymbol{R}^{(t)} \rangle + \alpha_t^2 \|\boldsymbol{R}^{(t)}\|_F^2} &(3.69b)\\ &\xrightarrow{\text{a.s.}} \frac{\|\boldsymbol{Z}^{(t)}\|_F^2 - \alpha_t(n + v_t n)}{\|\boldsymbol{Z}^{(t)}\|_F^2 - 2\alpha_t \|\boldsymbol{Z}^{(t)}\|_F^2 + \alpha_t^2(n + v_t n)} &(3.69c)\\ &= \frac{n + \lambda(1 + \frac{1}{\rho})nv_t + \lambda n v_t^2 \frac{1}{r} \sum_{i=1}^{r} \frac{1}{\theta_i^2} - \alpha_t(n + v_t n)}{(1 - 2\alpha_t)(n + \lambda(1 + \frac{1}{\rho})nv_t + \lambda n v_t^2 \frac{1}{r} \sum_{i=1}^{r} \frac{1}{\theta_i^2}) + \alpha_t^2(n + v_t n)} &(3.69d)\\ &= \frac{1 + \lambda(1 + \frac{1}{\rho})v_t + \lambda v_t^2 \frac{1}{r} \sum_{i=1}^{r} \frac{1}{\theta_i^2} - \alpha_t(1 + v_t)}{(1 - 2\alpha_t)(1 + \lambda(1 + \frac{1}{\rho})v_t + \lambda v_t^2 \frac{1}{r} \sum_{i=1}^{r} \frac{1}{\theta_i^2}) + \alpha_t^2(1 + v_t)} &(3.69e)\\ &\xrightarrow{\text{a.s.}} \frac{1 + \lambda(1 + \frac{1}{\rho})v_t + \lambda v_t^2 \Delta_2 - \alpha(v_t)(1 + v_t)}{(1 - 2\alpha(v_t))(1 + \lambda(1 + \frac{1}{\rho})v_t + \lambda v_t^2 \Delta_2) + (\alpha(v_t))^2(1 + v_t)} &(3.69f)\\ &= c(v_t) &(3.69g) \end{aligned}$$

where (3.69a) is from (3.10c), (3.69c) follows from Assumption 3.2 that $\boldsymbol{R}^{(t)} = \boldsymbol{X}_0 + \sqrt{v_t}\boldsymbol{N}$ with $\|\boldsymbol{X}_0\|_F^2 = n$ and the elements of $\boldsymbol{N}$ independently drawn from $\mathcal{N}(0, 1)$, (3.69d) is from (3.66), and (3.69f) is from the definition of Δ_2.

Appendix 5: Proof of Lemma 3.3

Assume that $\boldsymbol{R}^{(t)} = \sum_{i=1}^{\min(m,n)} \sigma_i \boldsymbol{u}_i \boldsymbol{v}_i^T$, where σ_i, $\boldsymbol{u}_i$, and $\boldsymbol{v}_i$ are the i-th singular value, left singular vector and right singular vectors, respectively. Then, $\boldsymbol{Z}^{(t)} = \sum_{i=1}^{r} \sigma_i \boldsymbol{u}_i \boldsymbol{v}_i^T$ since that $\boldsymbol{Z}^{(t)}$ is the best rank-r approximation of $\boldsymbol{R}^{(t)}$, and

$$\left\langle \boldsymbol{R}^{(t)} - \boldsymbol{Z}^{(t)}, \boldsymbol{Z}^{(t)} \right\rangle = \left\langle \sum_{j=r+1}^{\min(m,n)} \sigma_j \boldsymbol{u}_j \boldsymbol{v}_j^T, \sum_{i=1}^{r} \sigma_i \boldsymbol{u}_i \boldsymbol{v}_i^T \right\rangle \tag{3.70a}$$

$$= \sum_{i=1}^{r} \sum_{j=r+1}^{\min(m,n)} \sigma_i \sigma_j \left\langle \boldsymbol{u}_i \boldsymbol{v}_i^T, \boldsymbol{u}_j \boldsymbol{v}_j^T \right\rangle \tag{3.70b}$$

$$= \sum_{i=1}^{r} \sum_{j=r+1}^{\min(m,n)} \sigma_i \sigma_j \mathrm{Tr}(\boldsymbol{v}_i \boldsymbol{u}_i^T \boldsymbol{u}_j \boldsymbol{v}_j^T) \tag{3.70c}$$

$$= 0 \tag{3.70d}$$

where (3.70d) follows from that $\boldsymbol{u}_i^T \boldsymbol{u}_j = 0, \forall i, j$, and $i \neq j$. Recall from (3.41) that:

$$\left\langle \boldsymbol{R}^{(t)} - \boldsymbol{X}^{(t)}, \boldsymbol{X}^{(t)} \right\rangle = 0. \tag{3.71}$$

With the above orthogonalities, we have

$$\|\boldsymbol{X}^{(t)} - \boldsymbol{R}^{(t)}\|_F^2$$

$$= \|\boldsymbol{R}^{(t)}\|_F^2 - \|\boldsymbol{X}^{(t)}\|_F^2 \tag{3.72a}$$

$$= \|\boldsymbol{R}^{(t)}\|_F^2 - \left\| c_t (\boldsymbol{Z}^{(t)} - \alpha_t \boldsymbol{R}^{(t)}) \right\|_F^2 \tag{3.72b}$$

$$= \|\boldsymbol{R}^{(t)}\|_F^2 - \frac{\left\langle \boldsymbol{Z}^{(t)} - \alpha_t \boldsymbol{R}^{(t)}, \boldsymbol{R}^{(t)} \right\rangle^2}{\|\boldsymbol{Z}^{(t)} - \alpha_t \boldsymbol{R}^{(t)}\|_F^2} \tag{3.72c}$$

$$= \frac{\|\boldsymbol{R}^{(t)}\|_F^2 \|\boldsymbol{Z}^{(t)}\|_F^2 - \|\boldsymbol{Z}^{(t)}\|_F^4}{\|\boldsymbol{Z}^{(t)} - \alpha_t \boldsymbol{R}^{(t)}\|_F^2} \tag{3.72d}$$

$$= \frac{\|\boldsymbol{R}^{(t)}\|_F^2 \|\boldsymbol{Z}^{(t)}\|_F^2 - \|\boldsymbol{Z}^{(t)}\|_F^4}{\|\boldsymbol{Z}^{(t)}\|_F^2 - 2\alpha_t \left\langle \boldsymbol{R}^{(t)}, \boldsymbol{Z}^{(t)} \right\rangle + \alpha_t^2 \|\boldsymbol{R}^{(t)}\|_F^2} \tag{3.72e}$$

$$= \frac{\|\boldsymbol{R}^{(t)}\|_F^2 \|\boldsymbol{Z}^{(t)}\|_F^2 - \|\boldsymbol{Z}^{(t)}\|_F^4}{\|\boldsymbol{Z}^{(t)}\|_F^2 - 2\alpha_t \|\boldsymbol{Z}^{(t)}\|_F^2 + \alpha_t^2 \|\boldsymbol{R}^{(t)}\|_F^2} \tag{3.72f}$$

$$= \frac{\|\boldsymbol{R}^{(t)}\|_F^2 - \|\boldsymbol{Z}^{(t)}\|_F^2}{1 - 2\alpha_t + \alpha_t^2 + \alpha_t^2 \frac{\|\boldsymbol{R}^{(t)}\|_F^2}{\|\boldsymbol{Z}^{(t)}\|_F^2} - \alpha_t^2} \tag{3.72g}$$

$$= \frac{\|\boldsymbol{R}^{(t)}\|_F^2 - \|\boldsymbol{Z}^{(t)}\|_F^2}{\frac{\|\boldsymbol{R}\|_F^2 - \|\boldsymbol{Z}^{(t)}\|_F^2}{\|\boldsymbol{Z}^{(t)}\|_F^2}\alpha_t^2 + (1-\alpha_t)^2} \tag{3.72h}$$

$$= \frac{\|\boldsymbol{R}^{(t)} - \boldsymbol{Z}^{(t)}\|_F^2}{\frac{\|\boldsymbol{R}\|_F^2 - \|\boldsymbol{Z}^{(t)}\|_F^2}{\|\boldsymbol{Z}^{(t)}\|_F^2}\alpha_t^2 + (1-\alpha_t)^2} \tag{3.72i}$$

where (3.72a) follows from (3.71), (3.72b) follows by substituting $\boldsymbol{X}^{(t)}$ in Line 5 of Algorithm 3, (3.72c) follows by substituting c_t in (3.10c), and ((3.72d)–(3.72i)) follow from (3.70). This concludes the proof of Lemma 3.3.

Appendix 6: Proof of Theorem 3.2

From Condition 2 in (3.8) and Assumption 3.2, we have[4]

$$\left\langle \boldsymbol{R}^{(t)} - \boldsymbol{X}_0, \boldsymbol{X}_0 \right\rangle = 0 \tag{3.73a}$$

$$\left\langle \boldsymbol{R}^{(t)} - \boldsymbol{X}_0, \boldsymbol{X}^{(t)} - \boldsymbol{X}_0 \right\rangle = 0. \tag{3.73b}$$

Then,

$$\begin{aligned} &\|\boldsymbol{X}^{(t)} - \boldsymbol{X}_0\|_F^2 \\ &= \|\boldsymbol{X}^{(t)} - \boldsymbol{R}^{(t)}\|_F^2 - 2\left\langle \boldsymbol{R}^{(t)} - \boldsymbol{X}^{(t)}, \boldsymbol{R}^{(t)} - \boldsymbol{X}_0 \right\rangle \\ &\quad + \|\boldsymbol{R}^{(t)} - \boldsymbol{X}_0\|_F^2 && (3.74a) \\ &= \|\boldsymbol{R}^{(t)} - \boldsymbol{X}^{(t)}\|_F^2 - \|\boldsymbol{R}^{(t)} - \boldsymbol{X}_0\|_F^2 && (3.74b) \\ &= \frac{\|\boldsymbol{R}^{(t)}\|_F^2 - \|\boldsymbol{Z}^{(t)}\|_F^2}{\frac{\|\boldsymbol{R}\|_F^2 - \|\boldsymbol{Z}^{(t)}\|_F^2}{\|\boldsymbol{Z}^{(t)}\|_F^2}\alpha_t^2 + (1-\alpha_t)^2} - \|\boldsymbol{R}^{(t)} - \boldsymbol{X}_0\|_F^2 && (3.74c) \end{aligned}$$

[4]In fact, as $n_1, n_2, r \to \infty$ with $\frac{n_1}{n_2} \to \rho$ and $\frac{r}{n_2} \to \lambda$, the approximation in (3.14) becomes accurate, i.e. $\alpha_t = \frac{1}{n}\mathrm{div}(\mathcal{D}(\boldsymbol{R}^{(t)}))$ asymptotically satisfies Condition 2. Thus, (3.73b) asymptotically holds.

$$\xrightarrow{\text{a.s.}} \frac{nv_t - \lambda(1+\frac{1}{\rho})nv_t - \lambda n v_t^2 \Delta_2}{\frac{v_t - \lambda(1+\frac{1}{\rho})v_t - \lambda v_t^2 \Delta_2}{1+\lambda(1+\frac{1}{\rho})v_t + \lambda v_t^2 \Delta_2}(\alpha(v_t))^2 + (1-\alpha(v_t))^2} - nv_t \tag{3.74d}$$

where (3.74b) is from (3.73b), (3.74c) follows from (3.72), and (3.74d) follows from (3.67) and (3.68) and Assumption 3.2. Therefore, (3.23) holds, which concludes the proof of Theorem 3.2.

References

1. P. Jain, P. Netrapalli, S. Sanghavi, Low-rank matrix completion using alternating minimization, in *Proc. of the Forty-fifth Annual ACM Symposium on Theory of Computing (STOC)*, Palo Alto, June 2013, pp. 665–674
2. P. Jain, R. Meka, I.S. Dhillon, Guaranteed rank minimization via singular value projection, in *Proc. of Advances in Neural Information Processing Systems (NeurIPS)*, Vancouver, Dec 2010, pp. 937–945
3. T. Blumensath, M.E. Davies, Iterative hard thresholding for compressed sensing. Appl. Comput. Harmon. Anal. **27**(3), 265–274 (2009)
4. J. Tanner, K. Wei, Normalized iterative hard thresholding for matrix completion. SIAM J. Sci. Comput. **35**(5), S104–S125 (2013)
5. K. Wei, J.-F. Cai, T.F. Chan, S. Leung, Guarantees of Riemannian optimization for low rank matrix recovery. SIAM J. Matrix Anal. Appl. **37**(3), 1198–1222 (2016)
6. J. Ma, X. Yuan, L. Ping, Turbo compressed sensing with partial DFT sensing matrix. IEEE Signal Process. Lett. **22**(2), 158–161 (2015)
7. Z. Xue, J. Ma, X. Yuan, Denoising-based turbo compressed sensing. IEEE Access **5**, 7193–7204 (2017)
8. C. Eckart, G. Young, The approximation of one matrix by another of lower rank. Psychometrika **1**(3), 211–218 (1936)
9. E.J. Candès, B. Recht, Exact matrix completion via convex optimization. Found. Comput. Math. **9**(6), 717 (2009)
10. F. Benaych-Georges, R.R. Nadakuditi, The singular values and vectors of low rank perturbations of large rectangular random matrices. J. Multivar. Anal. **111**, 120–135 (2012)
11. A.M. Tulino, S. Verdú et al., Random matrix theory and wireless communications. Found. Trends Commun. Inf. Theory **1**(1), 1–182 (2004)
12. J. Ma, L. Ping, Orthogonal AMP. IEEE Access **5**, 2020–2033 (2017)
13. S. Rangan, P. Schniter, A.K. Fletcher, Vector approximate message passing. IEEE Trans. Inf. Theory **64**(10), 6664–6684 (2019)
14. K. Takeuchi, Rigorous dynamics of expectation-propagation-based signal recovery from unitarily invariant measurements, in *Proc. of IEEE International Symposium on Information Theory (ISIT)*, Aachen, June 2017
15. M. Bayati, A. Montanari, The dynamics of message passing on dense graphs, with applications to compressed sensing. IEEE Trans. Inf. Theory **57**(2), 764–785 (2011)
16. C.M. Stein, Estimation of the mean of a multivariate normal distribution. Ann. Stat. **9**(6), 1135–1151 (1981)
17. A. Kyrillidis, V. Cevher, Matrix recipes for hard thresholding methods. J. Math. Imaging Vis. **48**(2), 235–265 (2014)
18. B. Vandereycken, Low-rank matrix completion by Riemannian optimization. SIAM J. Optim. **23**(2), 1214–1236 (2013)

19. C.A. Metzler, A. Maleki, R.G. Baraniuk, From denoising to compressed sensing. IEEE Trans. Inf. Theory **62**(9), 5117–5144 (2016)
20. Z. Wen, W. Yin, Y. Zhang, Solving a low-rank factorization model for matrix completion by a nonlinear successive over-relaxation algorithm. Math. Program. Comput. **4**(4), 333–361 (2012)
21. B. Xin, D. Wipf, Pushing the limits of affine rank minimization by adapting probabilistic PCA, in *Proc. of International Conference on Machine Learning*, Lille, July 2015, pp. 419–427
22. K. Mohan, M. Fazel, Iterative reweighted algorithms for matrix rank minimization. J. Mach. Learn. Res. **13**(1), 3441–3473 (2012)
23. E.J. Candes, C.A. Sing-Long, J.D. Trzasko, Unbiased risk estimates for singular value thresholding and spectral estimators. IEEE Trans. Signal Process. **61**(19), 4643–4657 (2013)

Chapter 4
Turbo Message Passing for Compressed Robust Principal Component Analysis

4.1 Problem Description

Consider a noisy measurement $\boldsymbol{y} \in \mathbb{R}^{m\times 1}$ modeled by

$$\boldsymbol{y} = \mathcal{A}(\boldsymbol{L} + \boldsymbol{S}) + \boldsymbol{n} \tag{4.1}$$

where $\boldsymbol{L} \in \mathbb{R}^{n_1\times n_2}$ is a rank-r matrix to be recovered, $\boldsymbol{S} \in \mathbb{R}^{n_1\times n_2}$ is a sparse matrix representing the gross noise with at most k non-zeros, and $\boldsymbol{n} \in \mathbb{R}^{m\times 1}$ an additive white Gaussian noise with mean zero and variance $\sigma^2\boldsymbol{I}$. To reliably recover $\boldsymbol{L}$ and $\boldsymbol{S}$ from $\boldsymbol{y}$, we consider the following optimization problem:

$$\underset{\boldsymbol{L},\boldsymbol{S}}{\text{minimize}}\ \|\boldsymbol{y} - \mathcal{A}(\boldsymbol{L} + \boldsymbol{S})\|_2 \tag{4.2a}$$

$$\text{subject to}\ \text{rank}(\boldsymbol{L}) \le r \tag{4.2b}$$

$$\|\boldsymbol{S}\|_0 \le k. \tag{4.2c}$$

The above problem is difficult to solve since the rank and l_0-norm constraints are non-convex. This problem is closely related to the compressed sensing problem for sparse signal recovery, as well as to the problem of low-rank matrix recovery. Specifically, if we discard the low-rank matrix $\boldsymbol{L}$ and the rank constraint, then (4.2) reduces to a sparse signal recovery problem; if we discard the sparse matrix $\boldsymbol{S}$ and the l_0-norm constraint, then (4.2) becomes a low-rank matrix recovery problem. In [1], the authors developed a greedy algorithm to solve (4.2) by combining the CoSaMP algorithm [3] for sparse signal recovery and the ADMiRA algorithm [2] for low-rank matrix recovery. In [4], the authors constructed an unconstrained convex problem by relaxing the rank and l_0-norm constraints to nuclear norm and l_1-norm constraints. However, these algorithms do not perform well, especially when

X. Yuan and Z. Xue, *Turbo Message Passing Algorithms for Structured Signal Recovery*, SpringerBriefs in Computer Science,
https://doi.org/10.1007/978-3-030-54762-2_4

the recovery condition is pushed towards its limit (such as when k/n is increase and m/n decrease, where $n = n_1 n_2$). In this chapter, we will develop a turbo-type message passing algorithm for the compressed robust PCA problem, as detailed in the following.

4.2 Turbo-Type Message Passing Algorithms

4.2.1 TMP Framework

We now present the TMP framework for the compressed RPCA problem. The structure of TMP is shown in Fig. 4.1. There are three modules involved in TMP. Module A is to provide linear estimates of both $\boldsymbol{S}$ and $\boldsymbol{L}$ by exploiting the linear measurement $\boldsymbol{y}$. Module B is to estimate $\boldsymbol{S}$ by exploiting the property that $\boldsymbol{S}$ is sparse. Module C is to estimate $\boldsymbol{L}$ by exploiting the property that $\boldsymbol{L}$ is a low-rank matrix. These three modules are executed iteratively to refine the estimates of $\boldsymbol{S}$ and $\boldsymbol{L}$. Compared with the Turbo-CS in [5], TMP includes an extra low-rank matrix estimation module. Compared with the turbo-type affine rank minimization (TARM) algorithm in [6], TMP has an extra sparse matrix estimation module. In this sense, TMP combines the ideas of both Turbo-CS and TARM to yield an efficient algorithm for compressed RPCA.

Denote by $\boldsymbol{S}_A^{pri}$ and $\boldsymbol{L}_A^{pri}$, respectively, the input estimates of $\boldsymbol{S}$ and $\boldsymbol{L}$ of Module A, and by $v_{A,S}^{pri}$ and $v_{A,L}^{pri}$ the corresponding estimation MSEs; similarly, denote by $\boldsymbol{S}_A^{ext}$ and $\boldsymbol{L}_A^{ext}$, respectively, the output estimates of $\boldsymbol{S}$ and $\boldsymbol{L}$ of Module A, and by $v_{A,S}^{ext}$ and $v_{A,L}^{ext}$ the corresponding estimation MSEs, where superscript "pri" represents prior message, and superscript "ext" represents extrinsic message[1]. Similar notations apply to Modules B and C. From Fig. 4.1, the outputs of Modules

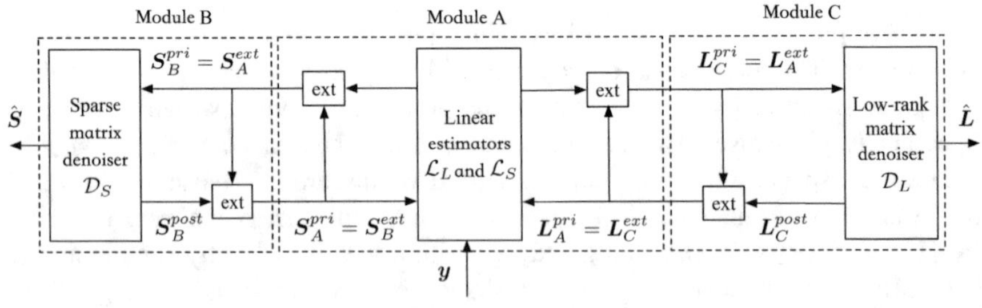

Fig. 4.1 The diagram of the TMP framework

[1]In message passing, a message of a variable node is a marginal distribution of this node. Particularly, for Gaussian message passing (as in TMP), a message refers to a pair of marginal mean and variance, since mean and variance uniquely determine a Gaussian distribution.

B and C are used as the inputs of Module A, and vice versa. That is, $\boldsymbol{S}_A^{pri} = \boldsymbol{S}_B^{ext}$, $\boldsymbol{L}_A^{pri} = \boldsymbol{L}_C^{ext}$, $\boldsymbol{S}_B^{pri} = \boldsymbol{S}_A^{ext}$, and $\boldsymbol{L}_C^{pri} = \boldsymbol{L}_A^{ext}$.

Similarly to D-Turbo-CS and TARM proposed in Chaps. 2 and 3, the calculation of extrinsic messages plays an essential role in TMP. Specifically, two conditions need to be met for the construction of the output of each module:

- **Condition 4.1**: *The output estimation error of a module is uncorrelated with the input estimation error of the module.*
- **Condition 4.2**: *The mean square error of the output of each module is minimized up to linear scaling.*

The above conditions are directly from the turbo message passing principle.

Conditions 4.1 and 4.2 are generally difficult to meet exactly in the algorithm design, since the statistical averages in the two conditions require the knowledge of the prior distributions of $\boldsymbol{S}$ and $\boldsymbol{L}$ (which is unfortunately difficult to acquire in practice). Even if the prior distributions are available, the integrals involved in the calculation may be difficult to evaluate. In the following subsections, we will provide details on how to (approximately) satisfy these two conditions at each module.

4.2.2 Module A: Linear Estimation

We now present the details of Module A. The input of Module A includes the linear measurement $\boldsymbol{y}$, the input estimate $\boldsymbol{S}_A^{pri}$, and the corresponding MSE $v_{A,S}^{pri}$ of $\boldsymbol{S}$ from Module B, and the input estimate $\boldsymbol{L}_A^{pri}$ and the corresponding MSE $v_{A,L}^{pri}$ of $\boldsymbol{L}$ from Module C. In Module A, we treat $\text{vec}(\boldsymbol{S}_A^{pri})$ and $v_{A,S}^{pri}\boldsymbol{I}$ as the prior mean and covariance of $\text{vec}(\boldsymbol{S})$, and treat $\text{vec}(\boldsymbol{L}_A^{pri})$ and $v_{A,L}^{pri}\boldsymbol{I}$ as the prior mean and covariance of $\text{vec}(\boldsymbol{L})$. Then, the linear minimum mean square error (MMSE) estimates of $\boldsymbol{S}$ and $\boldsymbol{L}$ given $\boldsymbol{y}$ can be, respectively, expressed as

$$\boldsymbol{S}_A^{post} = \boldsymbol{S}_A^{pri} + v_{A,S}^{pri}\mathcal{A}^T(((v_{A,L}^{pri} + v_{A,S}^{pri})\mathcal{A}\mathcal{A}^T + \sigma^2\mathcal{I})^{-1}(\boldsymbol{y} - \mathcal{A}(\boldsymbol{L}_A^{pri} + \boldsymbol{S}_A^{pri}))) \tag{4.3a}$$

$$\boldsymbol{L}_A^{post} = \boldsymbol{L}_A^{pri} + v_{A,L}^{pri}\mathcal{A}^T(((v_{A,L}^{pri} + v_{A,S}^{pri})\mathcal{A}\mathcal{A}^T + \sigma^2\mathcal{I})^{-1}(\boldsymbol{y} - \mathcal{A}(\boldsymbol{L}_A^{pri} + \boldsymbol{S}_A^{pri}))) \tag{4.3b}$$

with the corresponding MMSE matrices of $\text{vec}(\boldsymbol{S})$ and $\text{vec}(\boldsymbol{L})$ given by

$$\boldsymbol{V}_{A,S}^{post} = v_{A,S}^{pri}\boldsymbol{I} - (v_{A,S}^{pri})^2\boldsymbol{A}^T((v_{A,S}^{pri} + v_{A,L}^{pri})\boldsymbol{A}\boldsymbol{A}^T + \sigma^2\boldsymbol{I})^{-1}\boldsymbol{A} \tag{4.4a}$$

$$\boldsymbol{V}_{A,L}^{post} = v_{A,L}^{pri}\boldsymbol{I} - (v_{A,L}^{pri})^2\boldsymbol{A}^T((v_{A,S}^{pri} + v_{A,L}^{pri})\boldsymbol{A}\boldsymbol{A}^T + \sigma^2\boldsymbol{I})^{-1}\boldsymbol{A} \tag{4.4b}$$

where $\boldsymbol{A} \in \mathbb{R}^{m \times n}$ is the matrix form of linear operator $\mathcal{A}$. In (4.3) and (4.4), the superscript "^{post}" means posterior message. The posterior messages will be used to calculate the extrinsic messages based on Conditions 4.1 and 4.2.

We now construct the extrinsic messages such that Conditions 4.1 and 4.2 are met. We assume that the input estimates $\boldsymbol{S}_A^{pri}$ and $\boldsymbol{L}_A^{pri}$ are, respectively, given by

$$\boldsymbol{S}_A^{pri} = \boldsymbol{S} + \boldsymbol{N}_{A,S}^{pri} \tag{4.5a}$$

$$\boldsymbol{L}_A^{pri} = \boldsymbol{L} + \boldsymbol{N}_{A,L}^{pri} \tag{4.5b}$$

where the elements of additive noise $\boldsymbol{N}_{A,S}^{pri}$ are independent and identically distributed (i.i.d.) with mean zero and variance $v_{A,S}^{pri}$, and the elements of additive noise $\boldsymbol{N}_{A,L}^{pri}$ are i.i.d. with mean zero and variance $v_{A,L}^{pri}$.

We model the extrinsic means of $\boldsymbol{L}$ and $\boldsymbol{S}$ as a linear combination of the prior mean and the posterior mean:

$$S_{A,i,j}^{ext} = c_{S,i,j}(S_{A,i,j}^{post} - \alpha_{S,i,j} S_{A,i,j}^{pri}) \tag{4.6a}$$

$$L_{A,i,j}^{ext} = c_{L,i,j}(L_{A,i,j}^{post} - \alpha_{L,i,j} L_{A,i,j}^{pri}) \tag{4.6b}$$

where $c_{S,i,j}$, $c_{L,i,j}$, $\alpha_{S,i,j}$, and $\alpha_{L,i,j}$ are constant coefficients. The above modeling of extrinsic mean is a generalization of the extrinsic mean in [5]; see Eqn. (11) of [5]. Similar modeling of extrinsic mean has been previously used in [7] and [6].

Then, Conditions 4.1 and 4.2 yield the following optimization problems:

$$\underset{c_{S,i,j},\alpha_{S,i,j}}{\text{minimize}}\ \mathrm{E}[(S_{A,i,j}^{ext} - S_{i,j})^2] \tag{4.7a}$$

$$\text{subject to } \mathrm{E}[(S_{A,i,j}^{ext} - S_{i,j})(S_{A,i,j}^{pri} - S_{i,j})] = 0, \text{ for } \forall i, j \tag{4.7b}$$

$$\underset{c_{L,i,j},\alpha_{L,i,j}}{\text{minimize}}\ \mathrm{E}[(L_{A,i,j}^{ext} - L_{i,j})^2] \tag{4.8a}$$

$$\text{subject to } \mathrm{E}[(L_{A,i,j}^{ext} - L_{i,j})(L_{A,i,j}^{pri} - L_{i,j})] = 0, \text{ for } \forall i, j \tag{4.8b}$$

where $S_{A,i,j}^{ext}$, $S_{A,i,j}^{pri}$, and $S_{i,j}$ are, respectively, the (i, j)-th entries of $\boldsymbol{S}_A^{ext}$, $\boldsymbol{S}_A^{pri}$, and $\boldsymbol{S}$; $L_{A,i,j}^{ext}$, $L_{A,i,j}^{pri}$, and $L_{i,j}$ are, respectively, the (i, j)-th entries of $\boldsymbol{L}_A^{ext}$, $\boldsymbol{L}_A^{pri}$, and $\boldsymbol{L}$. The expectations in (4.7) and (4.8) are taken over the joint distribution of $\boldsymbol{S}$, $\boldsymbol{L}$, $\boldsymbol{N}_{A,S}^{pri}$, and $\boldsymbol{N}_{A,L}^{pri}$. Note that from the model in (4.5), $\boldsymbol{N}_{A,S}^{pri}$ and $\boldsymbol{N}_{A,L}^{pri}$ are independent of $\boldsymbol{S}$ and $\boldsymbol{L}$. Although the expectations in (4.7) and (4.8) involve the knowledge of the distributions of $\boldsymbol{S}$ and $\boldsymbol{L}$, these distributions are actually not required in determining the values of the coefficients $\{\alpha_{S,i,j}, \alpha_{L,i,j}, c_{S,i,j}, c_{L,i,j}\}$, as stated in the following theorem.

Theorem 4.1 *Assume that* $\boldsymbol{S}_{\mathrm{A}}^{pri}$ *and* $\boldsymbol{L}_{\mathrm{A}}^{pri}$ *satisfy (4.5), and the extrinsic means* $\boldsymbol{S}_{\mathrm{A}}^{ext}$ *and* $\boldsymbol{L}_{\mathrm{A}}^{ext}$ *have the form in (4.6). Then, the optimal solutions of problems (4.7) and (4.8) are given by*

$$\alpha_{l,i,j} = \frac{v_{A,l,i,j}^{post}}{v_{A,l}^{pri}} \tag{4.9a}$$

$$c_{l,i,j} = \frac{v_{A,l}^{pri}}{v_{A,l}^{pri} - v_{A,l,i,j}^{post}}, l \in \{L, S\}. \tag{4.9b}$$

The corresponding MSEs for the (i, j)*-th elements of* $\mathbf{S}$ *and* $\mathbf{L}$ *are given by*

$$v_{A,S,i,j}^{ext} = \mathrm{E}[(S_{A,i,j}^{ext} - S_{i,j})^2] = \frac{v_{A,S}^{pri} v_{A,S,i,j}^{post}}{v_{A,S}^{pri} - v_{A,S,i,j}^{post}} \tag{4.10a}$$

$$v_{A,L,i,j}^{ext} = \mathrm{E}[(L_{A,i,j}^{ext} - L_{i,j})^2] = \frac{v_{A,L}^{pri} v_{A,L,i,j}^{post}}{v_{A,L}^{pri} - v_{A,L,i,j}^{post}}. \tag{4.10b}$$

The proof of Theorem 4.1 is given in Appendix 1. From Theorem 4.1, the output estimates and the corresponding MSEs of $\boldsymbol{S}$ and $\boldsymbol{L}$ are given by

$$S_{A,i,j}^{ext} = \frac{v_{A,S}^{pri}}{v_{A,S}^{pri} - v_{A,S,i,j}^{post}} \left(S_{A,i,j}^{post} - \frac{v_{A,S,i,j}^{post}}{v_{A,S}^{pri}} S_{A,i,j}^{pri} \right) \tag{4.11a}$$

$$L_{A,i,j}^{ext} = \frac{v_{A,L}^{pri}}{v_{A,L}^{pri} - v_{A,S,i,j}^{post}} \left(L_{A,i,j}^{post} - \frac{v_{A,L,i,j}^{post}}{v_{A,L}^{pri}} L_{A,i,j}^{pri} \right) \tag{4.11b}$$

$$v_{A,l}^{ext} = \frac{1}{n} \sum_{i=1}^{n_1} \sum_{j=1}^{n_2} v_{A,l,i,j}^{ext}, l \in \{S, L\}, \tag{4.11c}$$

where $n = n_1 n_2$.

We next show that the calculations in (4.3) and (4.4) can be simplified when $\mathcal{A}$ is a partial orthogonal linear operator such that $\mathcal{A}\mathcal{A}^T = \mathcal{I}$. In this case, the LMMSE estimators in (4.3) have a concise form given by

$$\boldsymbol{S}_{A}^{post} = \boldsymbol{S}_{A}^{pri} + \frac{v_{A,S}^{pri}}{v_{A,L}^{pri} + v_{A,S}^{pri} + \sigma^2} \mathcal{A}^T(\boldsymbol{y} - \mathcal{A}(\boldsymbol{L}_{A}^{pri} + \boldsymbol{S}_{A}^{pri})) \tag{4.12a}$$

$$\boldsymbol{L}_A^{post} = \boldsymbol{L}_A^{pri} + \frac{v_{A,L}^{pri}}{v_{A,L}^{pri} + v_{A,S}^{pri} + \sigma^2} \mathcal{A}^T (\boldsymbol{y} - \mathcal{A}(\boldsymbol{L}_A^{pri} + \boldsymbol{S}_A^{pri})) \tag{4.12b}$$

where the corresponding MSEs are given by

$$v_{A,S}^{post} = v_{A,S}^{pri} - \frac{m}{n} \frac{(v_{A,S}^{pri})^2}{v_{A,S}^{pri} + v_{A,L}^{pri} + \sigma^2} \tag{4.13a}$$

$$v_{A,L}^{post} = v_{A,L}^{pri} - \frac{m}{n} \frac{(v_{A,L}^{pri})^2}{v_{A,S}^{pri} + v_{A,L}^{pri} + \sigma^2}. \tag{4.13b}$$

Then, from (4.11), the extrinsic outputs of Module A are given by

$$\boldsymbol{S}_A^{ext} = \boldsymbol{S}_A^{pri} + \frac{n}{m} \mathcal{A}^T (\boldsymbol{y} - \mathcal{A}(\boldsymbol{L}_A^{pri} + \boldsymbol{S}_A^{pri})) \tag{4.14a}$$

$$\boldsymbol{L}_A^{ext} = \boldsymbol{L}_A^{pri} + \frac{n}{m} \mathcal{A}^T (\boldsymbol{y} - \mathcal{A}(\boldsymbol{L}_A^{pri} + \boldsymbol{S}_A^{pri})) \tag{4.14b}$$

with the corresponding MSEs given by

$$v_{A,S}^{ext} = \frac{n}{m}(v_{A,S}^{pri} + v_{A,L}^{pri} + \sigma^2) - v_{A,S}^{pri} \tag{4.15a}$$

$$v_{A,L}^{ext} = \frac{n}{m}(v_{A,S}^{pri} + v_{A,L}^{pri} + \sigma^2) - v_{A,L}^{pri}. \tag{4.15b}$$

Note that the expressions in (4.14) can be seen as an extension of the Line 2 of Algorithm 1 in [7] for the estimation of two matrices $\boldsymbol{L}$ and $\boldsymbol{S}$. Also note that for a general $\mathcal{A}$, (4.12) gives a low-complexity alternative to (4.3) at the expense of certain performance loss.

4.2.3 Module B: Sparse Matrix Estimation

Module B takes $\boldsymbol{S}_B^{pri} = \boldsymbol{S}_A^{ext}$ and $v_{B,S}^{pri} = v_{A,S}^{ext}$ as its input. Denote by $\mathcal{D}_S(\cdot)$ the sparse matrix denoiser employed in Module B. That is,

$$\boldsymbol{S}_B^{post} = \mathcal{D}_S(\boldsymbol{S}_B^{pri}; v_{B,S}^{pri}). \tag{4.16}$$

Similarly to (4.6), we construct the extrinsic mean of $\boldsymbol{S}$ as a linear combination of the input and the output of $\mathcal{D}_S(\cdot)$:

$$S_{B,i,j}^{ext} = c_{S,i,j}(S_{B,i,j}^{post} - \alpha_{S,i,j} S_{B,i,j}^{pri}), \forall i, j \tag{4.17}$$

where $c_{S,i,j}$ and $\alpha_{S,i,j}$ are the combination coefficients. The following assumption is introduced to determine the combination coefficients.

Assumption 4.1 The input of Module B is modeled as $\boldsymbol{S}_B^{pri} = \boldsymbol{S} + \boldsymbol{N}_B^{pri}$, where $\boldsymbol{N}_B^{pri}$ is independent of $\boldsymbol{S}$ and the entries of $\boldsymbol{N}_B^{pri}$ are i.i.d. Gaussian distributed with mean zero and variance $v_{B,S}^{pri}$.

We emphasize that Assumption 4.1 is in essence a Gaussian-message approximation that has been widely used in the derivation of message passing algorithms, such as approximate message passing (AMP) [8], Turbo-CS [5], BiG-AMP [9], and P-BiG-AMP [10]. In some situations (such as AMP and Turbo-CS), approximate can be satisfied asymptotically by using the central limit theorem [11–13]. Yet, in many other situations (such as BiG-AMP and P-BiG-AMP), these approximations are not very accurate. This inaccuracy may compromise the convergence capability of the algorithm, and damping techniques are usually used to mitigate this effect. For TMP, we show that this assumption allows to calculate the extrinsic output of Module B based on Stein's lemma [14], as detailed below. We will discuss the accuracy of this assumption later in Sect. 4.5.

To satisfy Conditions 4.1 and 4.2, we need to solve the following problem:

$$\underset{\alpha_{S,i,j}, c_{S,i,j}}{\text{minimize}} \ \mathrm{E}[(S_{B,i,j}^{ext} - S_{i,j})_F^2] \tag{4.18a}$$

$$\text{subject to } \mathrm{E}[(S_{B,i,j}^{ext} - S_{i,j})(S_{B,i,j}^{pri} - S_{i,j})] = 0, \forall i, j \tag{4.18b}$$

where the expectation is taken over the joint distribution of $\boldsymbol{S}$ and $\boldsymbol{N}_B^{pri}$.

Theorem 4.2 *Assume that $\boldsymbol{S}_B^{ext}$ is of the form in (4.17) and $\boldsymbol{S}_B^{pri}$ satisfies Assumption 4.1. Then the optimal solution of problem (4.18) is given by*

$$\alpha_{S,i,j} = \mathrm{E}\left[\frac{\partial \mathcal{D}_{S,i,j}(\boldsymbol{S}_B^{pri}; v_{B,S}^{pri})}{\partial S_{B,i,j}^{pri}}\right] \tag{4.19a}$$

$$c_{S,i,j} = \frac{\mathrm{E}\left[(S_{B,i,j}^{post} - \alpha_{S,i,j} S_{B,i,j}^{pri}) S_{B,i,j}^{pri}\right]}{\mathrm{E}\left[(S_{B,i,j}^{post} - \alpha_{S,i,j} S_{B,i,j}^{pri})^2\right]}, \forall i, j \tag{4.19b}$$

where $\mathcal{D}_{S,i,j}(\boldsymbol{S}_B^{pri}; v_{B,S}^{pri})$ denotes the (i, j)-th element of $\mathcal{D}_S(\boldsymbol{S}_B^{pri}; v_{B,S}^{pri})$.

The proof of Theorem 4.2 is given in Appendix 2. The expectations in (4.19) are difficult to evaluate since they involve the distribution of $\boldsymbol{S}$ (which is usually unavailable). In the algorithm, we approximate $\alpha_{S,i,j}$ and $c_{S,i,j}$, respectively, by using the sample averages as

$$\alpha_S = \frac{1}{n}\sum_{i=1}^{n_1}\sum_{j=1}^{n_2}\frac{\partial \mathcal{D}_{S,i,j}(\boldsymbol{S}_B^{pri}; v_{B,S}^{pri})}{\partial S_{B,i,j}^{pri}} = \frac{1}{n}\text{div}(\mathcal{D}_S(\boldsymbol{S}_B^{pri}; v_{B,S}^{pri})) \tag{4.20a}$$

$$c_S = \frac{\left\langle \boldsymbol{S}_B^{post} - \alpha_S \boldsymbol{S}_B^{pri}), \boldsymbol{S}_B^{pri} \right\rangle}{\left\| \boldsymbol{S}_B^{post} - \alpha_S \boldsymbol{S}_B^{pri} \right\|^2} \tag{4.20b}$$

where div$(\cdot)$ denotes the divergence.

So far, we have discussed the generation of the extrinsic mean of Module B. We still need to determine the corresponding MSE. As a matter of fact, the extrinsic MSE $v_{B,S}^{ext}$ is dependent on the choice of denoiser $\mathcal{D}_S(\cdot)$. Thus, the update of $v_{B,S}^{ext}$ will be discussed in detail later when a specific denoiser is employed in Module B.

4.2.4 Module C: Low-Rank Matrix Estimation

The operations in Module C basically follow those in Module B. Specifically, Module C takes $\boldsymbol{L}_C^{pri} = \boldsymbol{L}_A^{ext}$ and $v_{C,L}^{pri} = v_{A,L}^{ext}$ as its input. Denote by $\mathcal{D}_L(\cdot)$ the low-rank matrix denoiser employed in Module C. The output of the denoiser can be expressed as

$$\boldsymbol{L}_C^{post} = \mathcal{D}_L(\boldsymbol{L}_C^{pri}; v_{C,L}^{pri}). \tag{4.21}$$

The extrinsic mean of $\boldsymbol{L}$ is constructed as a linear combination of the input and the output of $\mathcal{D}_L(\cdot)$:

$$L_{C,i,j}^{ext} = c_{L,i,j}(L_{C,i,j}^{post} - \alpha_{L,i,j} L_{C,i,j}^{pri}), \forall i, j \tag{4.22}$$

where $c_{L,i,j}$ and $\alpha_{L,i,j}$ are the combination coefficients. Then, similarly to Assumption 4.1, we introduce the following assumption.

Assumption 4.2 The input of Module C is modeled as $\boldsymbol{L}_C^{pri} = \boldsymbol{L} + \boldsymbol{N}_C^{pri}$, where $\boldsymbol{N}_C^{pri}$ is independent of $\boldsymbol{L}$ and the entries of $\boldsymbol{N}_C^{pri}$ are i.i.d. Gaussian distributed with mean zero and variance $v_{C,L}^{pri}$.

Based on (4.22) and Assumption 4.2, we solve the following problem to meet Conditions 4.1 and 4.2:

$$\underset{\alpha_{L,i,j}, c_{L,i,j}}{\text{minimize}}\ \text{E}[(L_{C,i,j}^{ext} - L_{i,j})^2] \tag{4.23a}$$

$$\text{subject to}\ \text{E}[(L_{C,i,j}^{ext} - L_{i,j}, L_{C,i,j}^{pri} - L_{i,j})] = 0, \forall i, j. \tag{4.23b}$$

The solution to (4.23) is given in the theorem below.

Theorem 4.3 *Assume that $\boldsymbol{L}_B^{ext}$ is of the form in (4.22), and that $\boldsymbol{L}_B^{pri}$ satisfies Assumption 4.2. Then the optimal solution of problem (4.23) is given by*

$$\alpha_{L,i,j} = \mathrm{E}\left[\frac{\partial \mathcal{D}_{L,i,j}(\boldsymbol{L}_C^{pri}; v_{c,L}^{pri})}{\partial L_{C,i,j}^{pri}}\right] \tag{4.24a}$$

$$c_{L,i,j} = \frac{\mathrm{E}\left[(L_{C,i,j}^{post} - \alpha_{L,i,j} L_{C,i,j}^{pri}) L_{C,i,j}^{pri}\right]}{\mathrm{E}\left[(L_{C,i,j}^{post} - \alpha_{L,i,j} L_{C,i,j}^{pri})^2\right]} \tag{4.24b}$$

where $\mathcal{D}_{L,i,j}(\boldsymbol{L}_C^{pri}; v_{c,L}^{pri})$ denotes the (i, j)-th element of $\mathcal{D}_L(\boldsymbol{L}_C^{pri}; v_{c,L}^{pri})$.

The proof of Theorem 4.3 is similar to that of Theorem 4.2, and thus omitted. Again, in the algorithm design, we approximate $\alpha_{S,i,j}$ and $c_{L,i,j}$, respectively, by using the sample averages as

$$\alpha_L = \frac{1}{n}\sum_{i=1}^{n_1}\sum_{j=1}^{n_2} \frac{\partial \mathcal{D}_{L,i,j}(\boldsymbol{L}_C^{pri}; v_{C,L}^{pri})}{\partial L_{C,i,j}^{pri}} = \frac{1}{n}\mathrm{div}(\mathcal{D}_L(\boldsymbol{L}_C^{pri}; v_{C,L}^{pri})) \tag{4.25a}$$

$$c_L = \frac{\left\langle \boldsymbol{L}_C^{post} - \alpha_L \boldsymbol{L}_C^{pri}), \boldsymbol{L}_C^{pri}\right\rangle}{\left\| \boldsymbol{L}_C^{post} - \alpha_L \boldsymbol{L}_C^{pri}\right\|^2}. \tag{4.25b}$$

The extrinsic MSE $v_{C,L}^{ext}$ of Module C is dependent on the choice of denoiser $\mathcal{D}_L(\cdot)$. The update of $v_{C,L}^{ext}$ will be discussed for each possible choice of $\mathcal{D}_L(\cdot)$ later.

4.2.5 Overall Algorithm

Based on the above discussions, we summarize the TMP algorithm in Algorithm 4. Note that the choices of $\mathcal{D}_S(\cdot)$ and $\mathcal{D}_L(\cdot)$, together with the updates of $v_{B,S}^{ext}$ and $v_{C,L}^{pri}$, will be elaborated later in Sects. 4.3 and 4.4.

The initialization of $\boldsymbol{S}_A^{pri}$, $\boldsymbol{L}_A^{pri}$, $v_{A,S}^{pri}$, and $v_{A,L}^{pri}$ is described as follows. We initialize $\boldsymbol{S}_A^{pri} = \boldsymbol{0}$ and $\boldsymbol{L}_A^{pri} = \boldsymbol{0}$. $v_{A,S}^{pri}$ and $v_{A,L}^{pri}$ are initialized as

$$v_{A,S}^{pri} = v_S = \frac{1}{n}\mathrm{E}[\|\boldsymbol{S}\|_F^2] \tag{4.26a}$$

$$v_{A,L}^{pri} = v_L = \frac{1}{n}\mathrm{E}[\|\boldsymbol{L}\|_F^2]. \tag{4.26b}$$

Numerical experiments show that the convergence of the TMP algorithm is not very sensitive to the initialization of $v_{A,S}^{pri}$ and $v_{A,L}^{pri}$. For example, when $\mathrm{E}[\|\boldsymbol{S}\|_F^2]$ and $\mathrm{E}[\|\boldsymbol{L}\|_F^2]$ are unknown, we can simply initialize $v_{A,S}^{pri}$ and $v_{A,L}^{pri}$ as

$$v_{A,S}^{pri} = v_S = \frac{1}{2m}\|\boldsymbol{y}\|_2^2 \tag{4.27a}$$

$$v_{A,L}^{pri} = v_L = \frac{1}{2m}\|\boldsymbol{y}\|_2^2. \tag{4.27b}$$

Algorithm 4 TMP algorithm for compressed RPCA

Input: $\mathcal{A}$, $\boldsymbol{y}$, $\boldsymbol{S}_A^{pri} = \boldsymbol{0}$, $\boldsymbol{L}_A^{pri} = \boldsymbol{0}$, $v_{A,S}^{pri} = v_S$, $v_{A,L}^{pri} = v_L$, σ

1: **while** the stopping criterion is not met **do**

2: $\boldsymbol{S}_A^{post} = \boldsymbol{S}_A^{pri} + v_{A,S}^{pri}\mathcal{A}^T(((v_{A,L}^{pri} + v_{A,S}^{pri})\mathcal{A}\mathcal{A}^T + \sigma^2\mathcal{I})^{-1}(\boldsymbol{y} - \mathcal{A}(\boldsymbol{L}_A^{pri} + \boldsymbol{S}_A^{pri})))$ % Module A

3: $\boldsymbol{L}_A^{post} = \boldsymbol{L}_A^{pri} + v_{A,L}^{pri}\mathcal{A}^T(((v_{A,L}^{pri} + v_{A,S}^{pri})\mathcal{A}\mathcal{A}^T + \sigma^2\mathcal{I})^{-1}(\boldsymbol{y} - \mathcal{A}(\boldsymbol{L}_A^{pri} + \boldsymbol{S}_A^{pri})))$

4: $\boldsymbol{V}_{A,S}^{post} = v_{A,S}^{pri}\boldsymbol{I} - (v_{A,S}^{pri})^2\boldsymbol{A}^T((v_{A,L}^{pri} + v_{A,S}^{pri})\boldsymbol{A}\boldsymbol{A}^T + \sigma^2\boldsymbol{I})^{-1}\boldsymbol{A}$

5: $\boldsymbol{V}_{A,L}^{post} = v_{A,L}^{pri}\boldsymbol{I} - (v_{A,L}^{pri})^2\boldsymbol{A}^T((v_{A,L}^{pri} + v_{A,S}^{pri})\boldsymbol{A}\boldsymbol{A}^T + \sigma^2\boldsymbol{I})^{-1}\boldsymbol{A}$

6: $S_{A,i,j}^{ext} = \frac{v_{A,S}^{pri}}{v_{A,S}^{pri} - v_{A,S,i,j}^{post}}(S_{A,i,j}^{post} - \frac{v_{A,S,i,j}^{post}}{v_{A,S}^{pri}}S_{A,i,j}^{pri}), \forall i, j$

7: $v_{A,S,i,j}^{ext} = \frac{v_{A,S}^{pri}v_{A,S,i,j}^{post}}{v_{A,S}^{pri} - v_{A,S,i,j}^{post}}, \forall i, j$

8: $L_{A,i,j}^{ext} = \frac{v_{A,L}^{pri}}{v_{A,L}^{pri} - v_{A,L,i,j}^{post}}(L_{A,i,j}^{post} - \frac{v_{A,L,i,j}^{post}}{v_{A,L}^{pri}}L_{A,i,j}^{pri}), \forall i, j$

9: $v_{A,L,i,j}^{ext} = \frac{v_{A,L}^{pri}v_{A,L,i,j}^{post}}{v_{A,L}^{pri} - v_{A,L,i,j}^{post}}, \forall i, j$

10: $\boldsymbol{S}_B^{pri} = \boldsymbol{S}_A^{ext}$; $\boldsymbol{L}_C^{pri} = \boldsymbol{L}_A^{ext}$; $v_{B,S}^{pri} = v_{A,S}^{ext} = \frac{1}{n}\sum_{i,j} v_{A,S,i,j}^{ext}$; $v_{C,L}^{pri} = v_{A,L}^{ext} = \frac{1}{n}\sum_{i,j} v_{A,L,i,j}^{ext}$

11: $\boldsymbol{S}_B^{post} = \mathcal{D}_S(\boldsymbol{S}_B^{pri}; v_{B,S}^{pri})$ % Module B

12: $\boldsymbol{S}_B^{ext} = c_S(\boldsymbol{S}_B^{post} - \alpha_S\boldsymbol{S}_B^{pri})$

13: Update $v_{B,S}^{ext}$ based on $\boldsymbol{S}_B^{ext}$

14: $\boldsymbol{L}_C^{post} = \mathcal{D}_L(\boldsymbol{L}_C^{pri}; v_{C,L}^{pri})$ % Module C

15: $\boldsymbol{L}_C^{ext} = c_L(\boldsymbol{L}_C^{post} - \alpha_L\boldsymbol{L}_C^{pri})$

16: Update $v_{C,L}^{ext}$ based on $\boldsymbol{L}_C^{ext}$

17: $\boldsymbol{S}_A^{pri} = \boldsymbol{S}_B^{ext}$; $\boldsymbol{L}_A^{pri} = \boldsymbol{L}_C^{ext}$; $v_{A,S}^{pri} = v_{B,S}^{ext}$; $v_{A,L}^{pri} = v_{C,L}^{ext}$

18: **end while**

Output: $\boldsymbol{L}_C^{post}$, $\boldsymbol{S}_B^{post}$

4.3 Sparse Matrix Denoisers for Module B

In this section, we discuss the choice of the sparse matrix denoiser for Module B. We focus on two candidate denoisers: the soft-thresholding denoiser [15] and the SURE-LET denoiser [16]. We will discuss how to appropriately tune the parameters of the two denoisers in Module B.

4.3.1 Preliminaries

We first present two useful lemmas. Consider

$$\boldsymbol{X} = \boldsymbol{X}_0 + \tau \boldsymbol{W} \in \mathbb{R}^{n_1 \times n_2} \tag{4.28}$$

where $\boldsymbol{X}_0$ is the original signal, τ is the noise level, and the entries of $\boldsymbol{W}$ are Gaussian distributed with zero mean and unit variance.

Lemma 4.1 ([14, Theorem 1]) *If $\mathcal{D}(\cdot)$ is weakly differentiable and $\boldsymbol{X}$ is modeled by (4.28), then*

$$\frac{1}{n}\mathrm{E}[\|\mathcal{D}(\boldsymbol{X}) - \boldsymbol{X}_0\|_F^2] = \frac{1}{n}\mathrm{E}[\|\mathcal{D}(\boldsymbol{X}) - \boldsymbol{X}\|_F^2] + \frac{2}{n}\tau^2\mathrm{E}[div(\mathcal{D}(\boldsymbol{X}))] - \tau^2. \tag{4.29}$$

For an arbitrary denoiser $\mathcal{D}(\cdot)$, define $\mathcal{D}^{ext}(\boldsymbol{X}) = \mathcal{D}(\boldsymbol{X}) - \frac{\mathrm{E}[\mathrm{div}(\mathcal{D}(\boldsymbol{X}))]}{n}\boldsymbol{X}$, where the expectation is taken over $\boldsymbol{X}_0$ and $\boldsymbol{W}$.

Lemma 4.2 *If $\mathcal{D}(\boldsymbol{X})$ is weakly differentiable, then*

$$\mathrm{E}[div(\mathcal{D}^{ext}(\boldsymbol{X}))] = 0. \tag{4.30}$$

Note that Lemma 4.1 can be proven by using Stein's lemma, and Lemma 4.2 can be verified straightforwardly by substituting the definition of $\mathcal{D}^{ext}(\cdot)$.

4.3.2 Soft-Thresholding Denoiser

The soft-thresholding denoiser is widely used in the field of image denoising and compressed sensing. The soft-thresholding denoiser is defined as

$$\eta(x; \lambda) = \mathrm{sign}(x)(|x| - \lambda)_+ \tag{4.31}$$

where λ is the threshold parameter. The $\eta(\cdot)$ function can be applied to a matrix in an entry-by -entry manner, yielding

$$\mathcal{D}_S(\boldsymbol{S}_B^{pri}; \lambda) = \eta(\boldsymbol{S}_B^{pri}; \lambda). \tag{4.32}$$

The parameter λ is related to the input noise level $v_{B,S}^{pri}$. For example, λ is chosen as $\lambda = \gamma\sqrt{v_{B,S}^{pri}}$ in [15], where γ is a certain fixed number. We choose λ as follows. From (4.17) and (4.20), we obtain $\boldsymbol{S}_B^{ext} = \mathcal{D}_S^{ext}(\boldsymbol{S}_B^{pri}) = c_S(\mathcal{D}_S(\boldsymbol{S}_B^{pri}; \lambda) - \alpha_S \boldsymbol{S}_B^{pri})$. Then, the MSE of $\boldsymbol{S}_B^{ext}$ is given by

$$\frac{1}{n}\mathrm{E}[\|\boldsymbol{S}_B^{ext} - \boldsymbol{S}\|_F^2] = \frac{1}{n}\mathrm{E}[\|\mathcal{D}_S^{ext}(\boldsymbol{S}_B^{pri}; \lambda) - \boldsymbol{S}\|_F^2] \tag{4.33a}$$

$$= \frac{1}{n}\mathrm{E}[\|\mathcal{D}_S^{ext}(\boldsymbol{S}_B^{pri}; \lambda) - \boldsymbol{S}_B^{pri}\|_F^2] + \frac{2}{n}v_{B,S}^{pri}\mathrm{E}[\mathrm{div}(\mathcal{D}_S^{ext}(\boldsymbol{S}_B^{pri}; \lambda))] - v_{B,S}^{pri} \tag{4.33b}$$

$$= \frac{1}{n}\mathrm{E}[\|\mathcal{D}_S^{ext}(\boldsymbol{S}_B^{pri}; \lambda) - \boldsymbol{S}_B^{pri}\|_F^2] - v_{B,S}^{pri} \tag{4.33c}$$

where (4.33b) follows from Lemma 4.1, and (4.33c) from Lemma 4.2. From (4.33), we see that the minimization of the MSE is equivalent to the minimization of $\mathrm{E}[\|\mathcal{D}_S^{ext}(\boldsymbol{S}_B^{pri}; \lambda) - \boldsymbol{S}_B^{pri}\|_F^2]$. To avoid the expectation (which requires the prior distribution of $\boldsymbol{S}$), we choose to find λ that minimizes $\|\mathcal{D}_S^{ext}(\boldsymbol{S}_B^{pri}; \lambda) - \boldsymbol{S}_B^{pri}\|_F^2$ instead of directly minimizing $\mathrm{E}[\|\mathcal{D}_S^{ext}(\boldsymbol{S}_B^{pri}; \lambda) - \boldsymbol{S}_B^{pri}\|_F^2]$. This is similar to the SURE approach in [14]. In the algorithm, we employ an exhaustive grid search over interval $(0, 10\sqrt{v_{B,S}^{pri}})$ to find the best λ.

The divergence of the soft-thresholding denoiser is given by

$$\mathrm{div}(\eta(\boldsymbol{S}_B^{pri}) = \sum_{i=1}^{n_1}\sum_{j=1}^{j=n_2} 1((\boldsymbol{S}_B^{pri})_{i,j} > \lambda). \tag{4.34}$$

4.3.3 SURE-LET Denoiser

The SURE-LET denoiser $\boldsymbol{D}_S(\cdot)$ [16] can be generally expressed as

$$\mathcal{D}_S(\boldsymbol{S}; \boldsymbol{\theta}) = \sum_{k=1}^{K} \theta_k \Psi_k(\boldsymbol{S}) \tag{4.35}$$

where $\{\Psi_k, k = 1, 2, \cdots, K\}$ are a set of k kernel functions, and $\boldsymbol{\theta} = [\theta_1, \cdots, \theta_K]$ is a vector of weights to be optimized. An exemplary set of kernel functions $\{\Psi_k(\cdot)\}$ is chosen as (2.28)–(2.30). Note that $\Psi_k(S)$ for $k = 1, 2, 3$ can be applied in an entry-by-entry manner to matrix $\boldsymbol{S}$, yielding $\Psi_k(\boldsymbol{S})$ with the (i, j)-th elements given by $\Psi_k(S_{i,j})$. β_1 and β_2 are constants chosen based on the input noise power $v_{B,S}^{pri}$. We set $\beta_1 = 2\sqrt{v}$ and $\beta_2 = 4\sqrt{v}$ following the recommendation in [17].

We next discuss how to construct the extrinsic denoiser based on the SURE-LET denoiser in (4.35). From (4.17) and (4.20), the extrinsic denoiser is given by

$$\mathcal{D}_S^{ext}(\boldsymbol{S}_B^{pri}; \boldsymbol{\theta}) = c_S(\mathcal{D}_S(\boldsymbol{S}_B^{pri}; \boldsymbol{\theta}) - \alpha_S \boldsymbol{S}_B^{pri}). \tag{4.36}$$

In the above, α_S is given in (4.20a) and c_S is determined together with $\boldsymbol{\theta}$ as detailed below.

We aim to minimize $\|\mathcal{D}_S^{ext}(\boldsymbol{S}_B^{pri}; \boldsymbol{\theta}) - \boldsymbol{S}_B^{pri}\|_F^2$ over $\boldsymbol{\theta}$. Define

$$\boldsymbol{\Psi}(\boldsymbol{S}_B^{pri}) = [\text{vec}(\Psi_1(\boldsymbol{S}_B^{pri})), \cdots, \text{vec}(\Psi_K(\boldsymbol{S}_B^{pri}))] \in \mathbb{R}^{n \times K} \tag{4.37a}$$

$$\text{div}(\boldsymbol{\Psi}(\boldsymbol{S}_B^{pri})) = [\text{div}(\text{vec}(\Psi_1(\boldsymbol{S}_B^{pri}))), \cdots, \text{div}(\text{vec}(\Psi_K(\boldsymbol{S}_B^{pri})))]^T \in \mathbb{R}^{K \times 1}. \tag{4.37b}$$

Then

$$\begin{aligned} & \|\mathcal{D}_S^{ext}(\boldsymbol{S}_B^{pri}; \boldsymbol{\theta}) - \boldsymbol{S}_B^{pri}\|_F^2 \\ &= \left\| c_S \left(\boldsymbol{\Psi}(\boldsymbol{S}_B^{pri})\boldsymbol{\theta} - \frac{1}{n}(\text{div}(\boldsymbol{\Psi}(\boldsymbol{S}_B^{pri}))^T \boldsymbol{\theta})\text{vec}(\boldsymbol{S}_B^{pri}) \right) - \text{vec}(\boldsymbol{S}_B^{pri}) \right\|_2^2 && (4.38a) \\ &= \left\| \left(\boldsymbol{\Psi}(\boldsymbol{S}_B^{pri}) - \frac{1}{n}\text{vec}(\boldsymbol{S}_B^{pri})\text{div}(\boldsymbol{\Psi}(\boldsymbol{S}_B^{pri}))^T \right) c_S \boldsymbol{\theta} - \text{vec}(\boldsymbol{S}_B^{pri}) \right\|_2^2. && (4.38b) \end{aligned}$$

To minimize (4.38), the optimal $\tilde{\boldsymbol{\theta}} = c_S \boldsymbol{\theta}$ is given by

$$\tilde{\boldsymbol{\theta}} = (\boldsymbol{\phi}(\boldsymbol{S}_B^{pri})^T \boldsymbol{\phi}(\boldsymbol{S}_B^{pri}))^{-1} \boldsymbol{\phi}(\boldsymbol{S}_B^{pri})^T \text{vec}(\boldsymbol{S}_B^{pri}) \tag{4.39}$$

where $\boldsymbol{\phi}(\boldsymbol{S}_B^{pri}) = \boldsymbol{\Psi}(\boldsymbol{S}_B^{pri}) - \frac{1}{n}\text{vec}(\boldsymbol{S}_B^{pri})\text{div}(\boldsymbol{\Psi}(\boldsymbol{S}_B^{pri}))^T$.

4.3.4 Output MSE of Module B

So far, we have discussed how to construct an extrinsic denoiser for sparse matrix estimation. Note that $\mathcal{D}_S^{ext}(\boldsymbol{S}_B^{pri})$ can be used as the output estimate of $\boldsymbol{S}$ for Module B. From the discussion in Sect. 4.2, we still need to determine the MSE of $\mathcal{D}_S^{ext}(\boldsymbol{S}_B^{pri})$ as an estimate of $\boldsymbol{S}$. We present two options for the estimate of the MSE:

$$v_{B,S}^{ext} = \frac{1}{n}\|\mathcal{D}_S^{ext}(\boldsymbol{S}_B^{pri};\lambda) - \boldsymbol{S}_B^{pri}\|_F^2 - v_{B,S}^{pri} \tag{4.40a}$$

$$v_{B,S}^{ext} = \frac{1}{m}\|\boldsymbol{y} - \mathcal{A}(\boldsymbol{S}_B^{ext} + \boldsymbol{L}_C^{ext})\|_2^2 - v_{C,L}^{ext} - \sigma^2. \tag{4.40b}$$

From (4.33), (4.40a) gives a good estimate of $v_{B,S}^{ext}$ when Assumption 4.1 holds. However, as shown later in Sect. 4.5, Assumption 4.1 is not very accurate. Here we present an alternative estimate of $v_{B,S}^{ext}$ in (4.40b). Clearly, when the size of the problem is large, (4.40b) becomes accurate. We will show by numerical simulation that (4.40b) gives a more robust estimate of $v_{B,S}^{ext}$ than (4.40a) does. In addition, we note that (4.40b) depends on the output MSE $v_{C,L}^{ext}$ of Module C. We will discuss how to robustly estimate $v_{C,L}^{ext}$ in the next section.

4.4 Low-Rank Denoisers for Module C

Module C needs a denoiser for the estimation of low-rank matrix $\boldsymbol{L}$ as detailed below.

4.4.1 Singular-Value Soft-Thresholding Denoiser

The singular value soft-thresholding (SVST) denoiser is defined as [18]

$$\mathcal{D}_L(\boldsymbol{L}_C^{pri};\lambda) = \arg\min_{\boldsymbol{L}} \frac{1}{2}\|\boldsymbol{L} - \boldsymbol{L}_B^{pri}\|_F^2 + \lambda\|\boldsymbol{L}\|_*. \tag{4.41}$$

Let the SVD of $\boldsymbol{L}_C^{pri}$ be $\boldsymbol{L}_C^{pri} = \boldsymbol{U}\boldsymbol{\Sigma}\boldsymbol{V}^T$, where $\boldsymbol{U} \in \mathbb{R}^{n_1\times n_1}$ and $\boldsymbol{V} \in \mathbb{R}^{n_2\times n_2}$ are orthogonal matrices, and $\boldsymbol{\Sigma} = \mathrm{diag}\{\sigma_1, \sigma_2, \cdots, \sigma_{\min(n_1,n_2)}\} \in \mathbb{R}^{n_1\times n_2}$ is a diagonal matrix with the diagonal elements arranged in the descending order. Then, the SVST denoiser given in (4.41) has a closed-form expression given by

$$\mathcal{D}_L(\boldsymbol{L}_C^{pri};\lambda) = \sum_{i=1}^{\min(n_1,n_2)} (\sigma_i - \lambda)_+ \boldsymbol{u}_i \boldsymbol{v}_i^T \tag{4.42}$$

where $\boldsymbol{u}_i$ is the i-th column of $\boldsymbol{U}$, and $\boldsymbol{v}_i$ is the i-th column of $\boldsymbol{V}$. The SVST denoiser applies soft-thresholding denoising to the singular values of the input matrix.

The thresholding parameter λ of the SVST denoiser needs to be tuned carefully to achieve a good denoising performance. The divergence of the SVST denoiser $\mathcal{D}(\boldsymbol{L}_B^{pri};\lambda)$ is given in [18] as

$$\operatorname{div}(\mathcal{D}(\boldsymbol{L}_C^{pri};\lambda)) = |n_1 - n_2| \sum_{i=1}^{\min(n_1,n_2)} \left(1 - \frac{\lambda}{\sigma_i}\right)_+ + \sum_{i=1}^{\min(n_1,n_2)} 1(\sigma_i > \lambda) + 2 \sum_{i,j=1,i\neq j}^{\min(n_1,n_2)} \frac{\sigma_i(\sigma_i - \lambda)_+}{\sigma_i^2 - \sigma_j^2}. \tag{4.43}$$

Then, the parameter α_L in (4.25a) can be calculated readily based on (4.43).

From (4.22) and (25), we obtain $\boldsymbol{L}_C^{ext} = c_L(\boldsymbol{L}_C^{post} - \alpha_L \boldsymbol{L}_C^{pri})$. Then, following the argument below (4.33), we determine λ by minimizing

$$\begin{aligned}
&\|\boldsymbol{L}_C^{ext} - \boldsymbol{L}_C^{pri}\|_F^2 \\
=&\|c_L(\boldsymbol{L}_C^{post} - \alpha_L \boldsymbol{L}_C^{pri}) - \boldsymbol{L}_C^{pri}\|_F^2 && (4.44a)\\
=&\left\| \frac{\left\langle \boldsymbol{L}_C^{post} - \alpha_L \boldsymbol{L}_C^{pri}, \boldsymbol{L}_C^{pri}\right\rangle}{\|\boldsymbol{L}_C^{post} - \alpha_L \boldsymbol{L}_C^{pri}\|_F^2}(\boldsymbol{L}_C^{post} - \alpha_L \boldsymbol{L}_C^{pri}) - \boldsymbol{L}_C^{pri}\right\|_F^2 && (4.44b)\\
=&\|\boldsymbol{L}_C^{pri}\|_F^2 - \frac{\left\langle \boldsymbol{L}_C^{post} - \alpha_L \boldsymbol{L}_C^{pri}, \boldsymbol{L}_C^{pri}\right\rangle^2}{\|\boldsymbol{L}_C^{post} - \alpha_L \boldsymbol{L}_C^{pri}\|_F^2} && (4.44c)\\
=&\|\boldsymbol{L}_C^{pri}\|_F^2 - \frac{\left\langle \mathcal{D}_L(\boldsymbol{L}_C^{pri};\lambda) - \alpha_L \boldsymbol{L}_C^{pri}, \boldsymbol{L}_C^{pri}\right\rangle^2}{\|\mathcal{D}_L(\boldsymbol{L}_C^{pri};\lambda) - \alpha_L \boldsymbol{L}_C^{pri}\|_F^2} && (4.44d)
\end{aligned}$$

where (4.44b) follows from (4.25). In (4.44d), the first term is irrelevant to λ. Thus, the optimal λ is given by

$$\lambda^* = \arg\max_{\lambda} \frac{\left\langle \boldsymbol{L}_C^{post} - \alpha_L \boldsymbol{L}_C^{pri}, \boldsymbol{L}_C^{pri}\right\rangle^2}{\|\boldsymbol{L}_C^{post} - \alpha_L \boldsymbol{L}_C^{pri}\|_F^2}. \tag{4.45}$$

In the algorithm, an exhaustive grid search over $(0, \sigma_1)$ is used to find the best λ.

4.4.2 Singular-Value Hard-Thresholding Denoiser

The singular value hard-thresholding (SVHT) denoiser is defined as

$$\mathcal{D}_L(\boldsymbol{L}_C^{pri};\lambda) = \arg\min_{\boldsymbol{L}} \frac{1}{2}\|\boldsymbol{L} - \boldsymbol{L}_C^{pri}\|_F^2 + \lambda \cdot \text{rank}(\boldsymbol{L}) \tag{4.46}$$

with the closed-form expression given by

$$\mathcal{D}_L(\boldsymbol{L}_C^{pri};\lambda) = \sum_{i=1}^{\min(n_1,n_2)} H(\sigma_i;\lambda)\boldsymbol{u}_i\boldsymbol{v}_i^T \tag{4.47}$$

where $H(x;\lambda)$ is the hard-thresholding function defined as

$$H(x;\lambda) = \begin{cases} x, & x > \lambda \\ 0, & \text{otherwise.} \end{cases} \tag{4.48}$$

From [18], the divergence of the SVHT denoiser $\mathcal{D}(\boldsymbol{L}_B^{pri},\lambda)$ is given by

$$\begin{aligned} \text{div}(\mathcal{D}(\boldsymbol{L}_C^{pri};\lambda)) = |n_1 - n_2| &\sum_{i=1}^{\min(n_1,n_2)} \frac{H(\sigma_i;\lambda)}{\sigma_i} \\ &+ \sum_{i=1}^{\min(n_1,n_2)} 1(\sigma_i > \lambda) + 2 \sum_{i,j=1,i\neq j}^{\min(n_1,n_2)} \frac{\sigma_i H(\sigma_i;\lambda)}{\sigma_i^2 - \sigma_j^2}. \end{aligned} \tag{4.49}$$

Then, the parameter λ of the SVHT denoiser can be determined by solving (4.45).

4.4.3 Best Rank-r Denoiser

The best rank-r denoiser is defined by the following optimization problem:

$$\underset{\boldsymbol{L}}{\text{minimize}} \ \|\boldsymbol{L} - \boldsymbol{L}_C^{pri}\|_F^2 \tag{4.50a}$$

$$\text{subject to} \ \text{rank}(\boldsymbol{L}) \leq r. \tag{4.50b}$$

The closed-form solution to problem (4.50) is given by

$$\boldsymbol{L}^* = \mathcal{D}_L(\boldsymbol{L}_C^{pri}) = \boldsymbol{U}\boldsymbol{\Sigma}_r\boldsymbol{V}^T \tag{4.51}$$

where $\boldsymbol{\Sigma}_r \in \mathbb{R}^{n_1\times n_2}$ is a diagonal matrix with the first r diagonal elements being $\sigma_1, \sigma_2, \cdots, \sigma_r$ and the others being 0. The parameter r of the best rank-r denoiser is assumed to be known by the algorithm.

From [18], the divergence of the best rank-r denoiser is given by

$$\operatorname{div}(\mathcal{D}_L(\boldsymbol{L}_B^{pri})) = |n_1 - n_2|r + r^2 + 2\sum_{i=1}^{r}\sum_{j=r+1}^{\min(n_1,n_2)} \frac{\sigma_i^2}{\sigma_i^2 - \sigma_j^2}. \tag{4.52}$$

4.4.4 Output MSE of Module C

An accurate estimation of the output MSE of Module C is very important to the convergence of TMP. We now give three options for the estimation of the output MSE of Module C:

$$v_{C,L}^{ext} = \frac{\|\boldsymbol{L}_C^{pri}\|_F^2}{n} - \frac{\left\langle \mathcal{D}_L(\boldsymbol{L}_C^{pri};\lambda) - \alpha_L \boldsymbol{L}_C^{pri}, \boldsymbol{L}_C^{pri}\right\rangle^2}{n\|\mathcal{D}_L(\boldsymbol{L}_C^{pri};\lambda) - \alpha_L \boldsymbol{L}_C^{pri}\|_F^2} - v_{C,L}^{pri} \tag{4.53a}$$

$$v_{C,L}^{ext} = \frac{\|\boldsymbol{L}_C^{ext}\|_F^2}{\|\boldsymbol{L}_C^{pri}\|_F^2} v_{C,L}^{pri} \tag{4.53b}$$

$$v_{C,L}^{ext} = v_{B,L}^{pri}\left(\left(1 - \frac{r}{n_2}\left(1 + \frac{n_2}{n_1}\right)\right)\frac{1}{(1-\alpha_L)^2} - 1\right). \tag{4.53c}$$

From (4.44), (4.53a) gives a good estimate under Assumption 4.2. However, as seen later in Sect. 4.5, Assumption 4.2 is not very accurate. Thus, we propose to use (4.53b) for a more robust estimation of the output MSE of Module C. It is also worth noting that for best-rank r denoiser, we can use (4.53c) for the MSE estimation [6]. Later in Sect. 4.6, we will give detailed comparisons of TMP with different denoisers and MSE estimations.

4.5 Convergence and Complexity of TMP

4.5.1 Revisiting Assumptions 4.1 and 4.2

Recall that Assumptions 4.1 and 4.2 are introduced to decouple the probability space of the modules and simplify the derivation of TMP. We next check the validity of these assumptions by numerical simulations.

Since Assumptions 4.1 and 4.2 are similar, we take Assumption 4.2 as an example and numerically verify the Gaussianity of $\boldsymbol{N}_{C,L}^{pri}$. For convenience of discussion, we choose the linear operator $\mathcal{A}$ to be partial orthogonal. Then

$$\boldsymbol{N}_{C,L}^{pri}$$

$$= \boldsymbol{L}_C^{pri} - \boldsymbol{L} = \boldsymbol{L}_A^{ext} - \boldsymbol{L} \tag{4.54a}$$

$$= \boldsymbol{L}_A^{pri} + \frac{n}{m}\mathcal{A}^T(\boldsymbol{y} - \mathcal{A}(\boldsymbol{L}_A^{pri} + \boldsymbol{S}_A^{pri})) - \boldsymbol{L} \tag{4.54b}$$

$$= (\mathcal{I} - \frac{n}{m}\mathcal{A}^T\mathcal{A})(\boldsymbol{L}_A^{pri} - \boldsymbol{L}) - \frac{n}{m}\mathcal{A}^T\mathcal{A}(\boldsymbol{S}_A^{pri} - \boldsymbol{S}) + \frac{n}{m}\mathcal{A}^T(\boldsymbol{n}) \tag{4.54c}$$

where (4.54c) is from (4.14b). From (4.54c), we see that the distortion $\boldsymbol{N}_{C,L}^{pri}$ consists of three terms. The third term is Gaussian since a linear transform of the Gaussian noise $\boldsymbol{n}$ is Gaussian. It is interesting to note that the first term also occurs in the input error of the low-rank matrix denoiser in the TARM algorithm. It was shown in [6] that this term can be well approximated as i.i.d. Gaussian distributed. The second term in (4.54c) is due to the existence of two matrices $\boldsymbol{L}$ and $\boldsymbol{S}$ in the linear estimation.

The QQ plots of the first and second terms of (4.54c) are, respectively, plotted in Figs. 4.2 and 4.3. From Fig. 4.2, we see that the first term of (4.54c) is very close to Gaussian throughout the iterative process; however, from Fig. 4.3, we see that the second term of (4.54c) deviates from Gaussian to some extent in the iterative process. Furthermore, such deviation becomes more serious when the measurement rate $\frac{m}{n}$ is close to 1.

The above deviation from Gaussianity has two consequences to the TMP algorithm. First, the variance evolution of TMP is not accurate. Second, the convergence of the TMP algorithm may be compromised when such deviation is serious. As such,

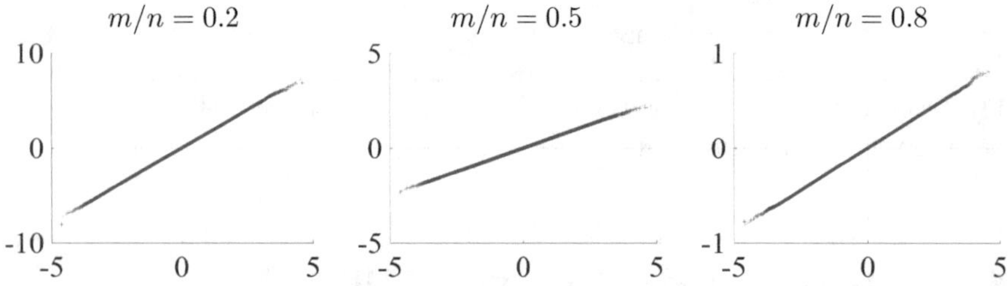

Fig. 4.2 The QQ plot of the first part of estimation error of $\boldsymbol{L}$ in Module A given in (4.54b) at the second iteration of TMP. The settings are $n_1 = n_2 = 500$, $r = 20$, $k/n = 0.1$, and $\sigma^2 = 10^{-5}$

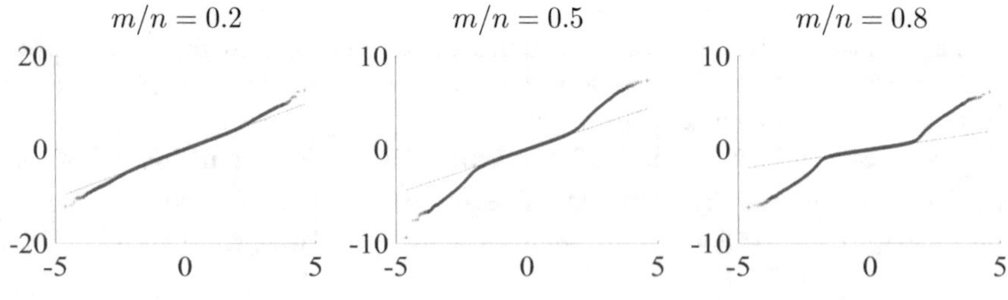

Fig. 4.3 The QQ plot of the second part of estimation error of $\boldsymbol{L}$ in Module A given in (4.54b) at the second iteration of TMP. The settings are $n_1 = n_2 = 500$, $r = 20$, $k/n = 0.1$, and $\sigma^2 = 10^{-5}$

the damping technique is employed in TMP to improve convergence, as detailed below.

4.5.2 Damping Technique

To improve convergence of the TMP algorithm, we damp the output of Module A as

$$\boldsymbol{L}_{A,t}^{ext} = \beta \boldsymbol{L}_{A,t}^{ext} + (1-\beta)\boldsymbol{L}_{A,t-1}^{ext} \tag{4.55}$$

$$\boldsymbol{S}_{A,t}^{ext} = \beta \boldsymbol{S}_{A,t}^{ext} + (1-\beta)\boldsymbol{S}_{A,t-1}^{ext} \tag{4.56}$$

where $\beta \in (0, 1]$ is the damping factor, and the subscript "$_t$" represents the t-th iteration. Similarly, we damp the MSEs as

$$v_{A,S,t}^{ext} = \beta v_{A,S,t}^{ext} + (1-\beta) v_{A,S,t-1}^{ext} \tag{4.57}$$

$$v_{A,L,t}^{ext} = \beta v_{A,L,t}^{ext} + (1-\beta) v_{A,L,t-1}^{ext}. \tag{4.58}$$

The choice of the damping factor β is crucial to improve convergence of TMP. We have two choices of β. The first option is to set β as a predetermined constant in the algorithm. The second option is to choose β adaptively by following the approach in [19].

For the second option, we follow [19] to construct a cost function as

$$\begin{aligned}\hat{J} = &\sum_{i=1}^{n_1}\sum_{j=1}^{n_2} D(\mathcal{N}(L_{i,j}; L_{A,i,j}^{ext}, v_{A,L}^{ext}) \| \mathcal{N}(L_{i,j}; 0, v_L)) + \\ &\sum_{i=1}^{n_1}\sum_{j=1}^{n_2} D(\mathcal{N}(S_{i,j}; S_{A,i,j}^{ext}, v_{A,S}^{ext}) \| \mathcal{N}(S_{i,j}; 0, v_S)) - \\ &\sum_{i=1}^{m} \mathrm{e}_{z_i \sim \mathcal{N}(z_i; z_i^{ext}, v_z^{ext})} \log(\mathcal{N}(z_i; y_i, \sigma^2)))\end{aligned} \tag{4.59}$$

where $D(\cdot\|\cdot)$ denotes the KL-divergence, and z^{ext} and v_z^{ext} are given by

$$z^{ext} = \mathcal{A}(\boldsymbol{L}_A^{ext} + \boldsymbol{S}_A^{ext}) \tag{4.60a}$$

$$v_z^{ext} = v_{A,S}^{ext} + v_{A,L}^{ext}. \tag{4.60b}$$

We set the initial value of β to 0.5. At each iteration, we calculate the cost function $\hat{J}$ which is expected to decrease at each iteration. When $\hat{J}$ indeed decreases, we set $\beta = \max(0.95\beta, 0.05)$; otherwise, set $\beta = \min(1.3\beta, 0.95)$.

4.5.3 Complexity of TMP

We now briefly discuss the computational complexity involved in the TMP algorithm. Each iteration of TMP involves the execution of three modules. For module A, the complexity is dominated by the matrix inversion, with worst-case complexity $O(m^3)$. This complexity can be reduced to $O(mn_1n_2)$ if the matrix form $\boldsymbol{A}$ of the linear operator is pre-factorized using SVD (by following the approach in Algorithm 2 of [13]). For partial orthogonal linear operators, the complexity of lines 1-4 of Algorithm 1 is still $O(mn_1n_2)$, but the pre-factorization of $\boldsymbol{A}$ is not necessary any more. The complexity can be further reduced to $O(m\log(n_1n_2))$ if the measurement matrix $\boldsymbol{A}$ is a partial-DFT or partial-DCT matrix.

Module B involves only entry-by-entry calculation with marginal complexity. The parameter tuning of the SURE-LET denoiser in (4.39) involves the calculation of the inverse of a $K \times K$ matrix. Typically, K is small, and so the complexity is marginal.

Module C involves the calculation of the SVD of $\boldsymbol{L}_C^{pri}$ with complexity $O(n_1^2n_2)$. When the best rank-r denoiser is employed, we only need to calculate the largest r singular values and the corresponding singular vectors with complexity $O(rn_1n_2)$.

It is also interesting to compare the complexity of TMP with that of P-BiG-AMP. P-BiG-AMP does not require matrix inversion and SVDs. However, it requires matrix multiplications and thus has comparable per-iteration complexity as TMP. For example, the complexity of R1, R2, and R5 of P-BiG-AMP (in Table IV of [10]) is $O(mn_1n_2r^2)$ for a general linear operator. Although TMP and P-BiG-AMP have comparable per-iteration complexity, the running time of TMP is usually much shorter than that of P-BiG-AMP due to the following two reasons. First, TMP converges much faster than P-BiG-AMP, thanks to the fast convergence of the Turbo-CS algorithm and the avoidance of matrix factorization. Second, the performance of P-BiG-AMP is sensitive to initialization. To achieve a good performance, P-BiG-AMP needs to try a number of random initializations and pick the one with the best performance. From our numerical experiments, we find that such a try-and-error approach is very time consuming, especially under boundary conditions for successful recovery.

4.6 Numerical Results

In this section, we present numerical results of TMP for the compressed RPCA problem. We consider three types of linear operators $\mathcal{A}$: the random partial orthogonal linear operator, the random selection operator that randomly selects m entries from the input matrix, and the random Gaussian linear operator. When the linear operator is chosen as the random selection operator, the problem becomes a robust matrix completion problem.

In the following experiments, the rank-r matrix $\boldsymbol{L}$ is generated by the multiplication of two zero mean Gaussian random matrices of sizes $n_1 \times r$ and $r \times n_2$. We normalize $\boldsymbol{L}$ such that $\|\boldsymbol{L}\|_F^2 = n$. The elements of sparse matrix $\boldsymbol{S}$ are independently generated by following a Gaussian-Bernoulli distribution with zero mean and unit variance. The random partial orthogonal linear operator $\mathcal{A}$ is generated as

$$\mathcal{A}(\cdot) = \mathcal{S}(IDCT(Mask2(DCT(Mask1(\cdot))))) \tag{4.61}$$

where $Mask1(\cdot)$ and $Mask2(\cdot)$ are mask operators that randomly flip the sign of the entries of the input matrix, $DCT(\cdot)$ is a discrete cosine transform (DCT) operator, $IDCT(\cdot)$ is an inverse DCT operator, and $\mathcal{S}(\cdot)$ is a random selection linear operator. All the simulated curves are obtained by taking sample averages over 10 random realizations, unless otherwise specified.

4.6.1 TMP with Different Denoisers and MSE Estimations

In this subsection, we compare the performance of the TMP algorithm with the denoisers and the MSE estimators discussed in Sects. 4.3 and 4.4. The linear operator $\mathcal{A}$ is chosen as the random partial orthogonal linear operator in (4.61). In Fig. 4.4, we fix the sparse matrix denoiser to be the SURE-LET denoiser and choose $\boldsymbol{v}_{B,S}^{ext}$ as (4.40b) and compare the three different low-rank matrix denoisers presented in Sect. 4.3. As shown in Fig. 4.4, when the best rank-r denoiser is employed and $\boldsymbol{v}_{B,S}^{ext}$ is chosen as (4.53c), TMP has the fastest convergence; TMP with SVST denoiser converges with the slowest rate and has the worst MSE; TMP with SVHT

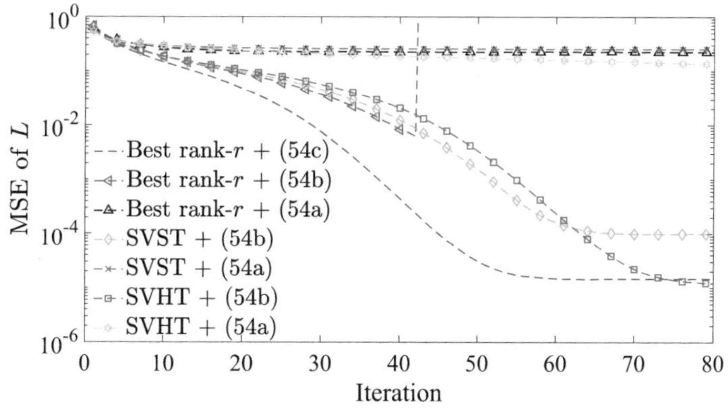

Fig. 4.4 Comparison of the TMP algorithms with different low-rank matrix denoisers. The sparse matrix denoiser is fixed as the SURE-LET denoiser. The other settings are $n_1 = n_2 = 250$, $m = 0.4n$, $r = 20$, $\sigma^2 = 10^{-5}$, $\|\boldsymbol{S}\|_0 = 0.1n$, and $\beta = 0.9$

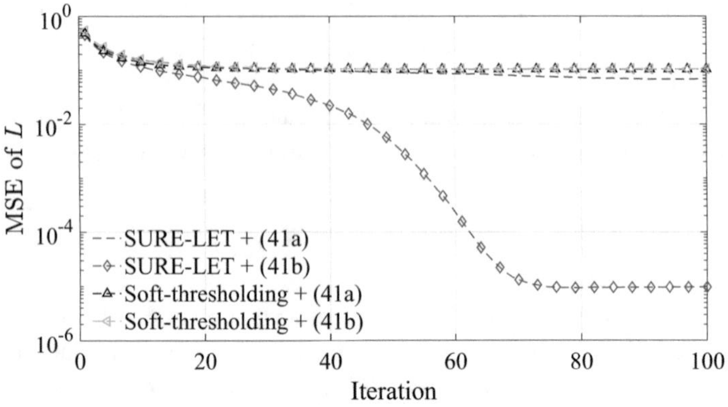

Fig. 4.5 Comparison of the TMP algorithms with different sparse matrix denoisers. The low-rank matrix denoiser is fixed as the best rank-r denoiser. The other settings are $n_1 = n_2 = 250$, $m = 0.3n$, $r = 10$, $\sigma^2 = 10^{-5}$, $\|\boldsymbol{S}\|_0 = 0.1n$, and $\beta = 0.9$

denoiser has the best MSE, while it converges slower than TMP with the best rank-r denoiser. When the best rank-r denoiser is employed and $\boldsymbol{v}_{B,S}^{ext}$ is chosen as (4.53b), TMP diverges. In Fig. 4.5, we compare TMP with different sparse denoisers. We fix the low-rank matrix denoiser to be the best rank-r denoiser and choose $\boldsymbol{v}_{C,L}^{ext}$ given by (4.53c). As shown in Fig. 4.5, when the SURE-LET denoiser is employed and $\boldsymbol{v}_{B,S}^{ext}$ is chosen as (4.40b), TMP has the fastest convergence rate and the lowest converged MSE.

4.6.2 *Convergence Comparison of TMP with Existing Algorithms*

In the following simulation, the sparse matrix denoiser and the low-rank matrix denoiser employed in TMP are always chosen as the SURE-LET denoiser and the best rank-r denoiser, respectively. The output MSE estimator for the SURE-LET denoiser is fixed to (4.40b), and that for the best rank-r denoiser is (4.53c). We compare TMP with SpaRCS [1], SPCP [20], and P-BiG-AMP [10] under three different choices of the linear operators.

4.6.2.1 Partial Orthogonal Linear Operator

The linear operator $\mathcal{A}$ is chosen as the partial orthogonal linear operator given in (4.61). Here the measurement rate $\frac{m}{n}$ is set relatively low, and a relatively small damping factor is used in the TMP algorithm. From Fig. 4.6, TMP has a much lower

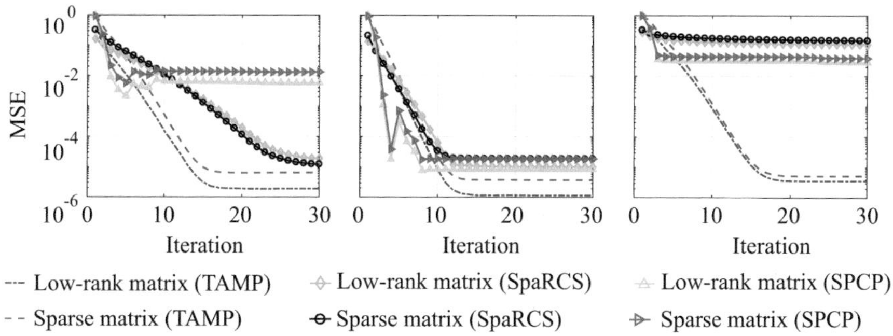

Fig. 4.6 Comparison of TMP with SpaRCS under random partial orthogonal linear operator. Left: The settings are $n_1 = n_2 = 256, m = 0.15n, r = 2, \sigma^2 = 10^{-5}$, and sparsity rate $\|\boldsymbol{S}\|_0 = 0.02n$. Middle: The settings are $n_1 = n_2 = 256, m = 0.2n, r = 2, \sigma^2 = 10^{-5}$, and sparsity rate $\|\boldsymbol{S}\|_0 = 0.02n$. Right: The settings are $n_1 = n_2 = 512, m = 0.5n, r = 25, \sigma^2 = 10^{-5}$, and sparsity rate $\|\boldsymbol{S}\|_0 = 0.1n$

Table 4.1 The running time comparisons of TMP and P-BiG-AMP for partial orthogonal linear operator. $n_1 = n_2 = 64$ and $\sigma^2 = 10^{-10}$

Parameters			TMP		P-BiG-AMP	
m/n	k/m	r	Time	NS	Time	NS
0.5	0.25	2	0.89	10	130.52	10
0.35	0.05	2	0.30	10	9.22	10
0.5	0.2	3	0.34	10	14.59	10
0.4	0.1	3	0.36	10	113.78	10
0.35	0.01	3	0.37	10	11.15	10

MSE than SpaRCS and SPCP in order of magnitude in most settings.[2] Empirically, we observe that the running time of TMP is only a fraction of that of SpaRCS, and that TMP is order-of-magnitude faster than SPCP. Note that in the middle plot of Fig. 4.6, SPCP has the least iteration times to converge. However, SPCP requires solving hundreds of constrained RPCA problem at each iteration which is far more complex than TMP and SpaRCS. We also observe that the performance of SPCP is not very stable, especially in some extreme conditions (such as in the left plot of Fig. 4.6). Therefore, we henceforth do not include SPCP in the performance comparison. In Table 4.1, we compare the running time of TMP with that of P-BiG-AMP when linear operator is partial orthogonal. A relatively small size of the problem, i.e., $n_1 = n_2 = 64$, is considered in simulation; NS in Table 4.1 denotes the number of successful recovery out of 10 random simulations. We see that the running time of TMP is at least one-order-of-magnitude shorter than that of P-BiG-AMP under all considered settings.

[2]P-BiG-AMP has not been included for comparison in Fig. 4.6, since it requires to store the whole sensing operator which is storage consuming.

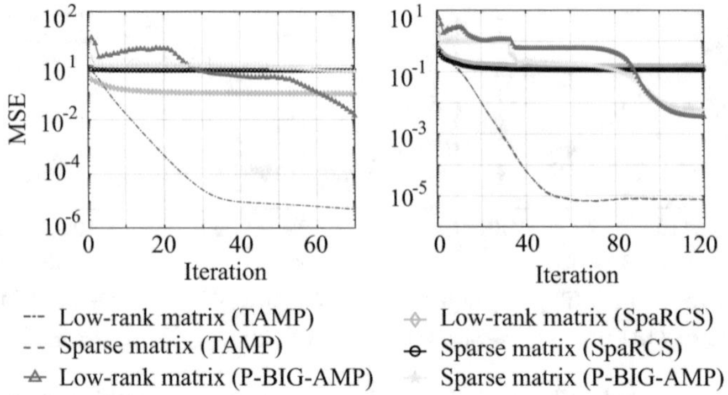

Fig. 4.7 Left: comparison of TMP with SpaRCS and P-BiG-AMP under random selection linear operator. The settings are $\frac{m}{n} = 0.4$, $n_1 = n_2 = 256, r = 10, \sigma^2 = 10^{-5}$, $\|S\|_0 = 0.1n$, and damping factor $\beta = 0.3$. Right: comparison of TMP with SpaRCS and P-BiG-AMP under random Gaussian linear operator. The settings are $\frac{m}{n} = 0.5$, $n_1 = n_2 = 80, r = 5, \sigma^2 = 10^{-5}$, $\|S\|_0 = 0.05n$, and damping factor $\beta = 0.5$

4.6.2.2 Robust Matrix Completion

We choose linear operator $\mathcal{A}$ as the random selection linear operator and compare TMP with SpaRCS in Fig. 4.7. The damping factor β is fixed at 0.3. From Fig. 4.7, we see that TMP has a much faster convergence rate and lower converged MSE than the counterpart algorithms.

4.6.2.3 Gaussian Linear Operator

In the right figure of Fig. 4.7, we choose linear operator $\mathcal{A}$ to be a random Gaussian linear operator. From Fig. 4.7, we see again that TMP has a much faster convergence rate and lower converged MSE than SpaRCS and P-BiG-AMP.

4.6.3 Phase Transition

The phase transition curve gives the boundary between the regions of successful/unsuccessful recovery of an algorithm. In simulations, the recovery is treated as successful when the normalized residue MSE $\frac{\|y-\mathcal{A}(\hat{L}+\hat{S})\|_2^2}{\|y\|_2^2} \leq 10^{-9}$. Figure 4.8 shows the phase transition curves of TMP with $n_1 = n_2 = 64$ for different values of r, k, and m. The x-axis of the plots in Fig. 4.8 is the measurement rate m/n, and the y-axis is the sparsity rate k/m. From [21], for successful recovery, the number of measurements should be no less than $(n_1 + n_2 - r)r + k$ (shown as the upper bound in Fig. 4.8). As expected, the performance of TMP degrades as the increase of k/m

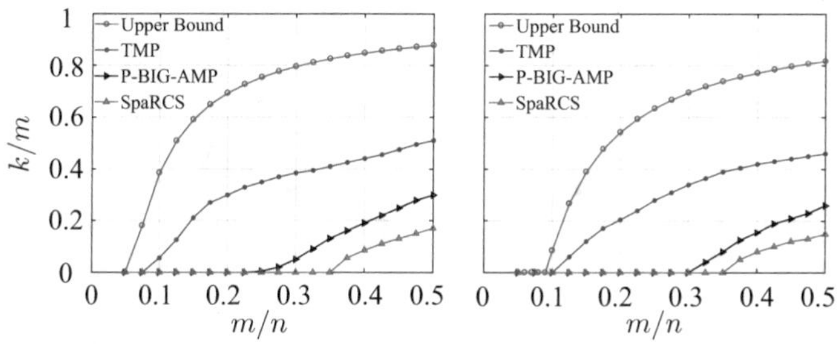

Fig. 4.8 Phase transitions of TMP, P-BiG-AMP, and SpaRCS under partial orthogonal linear operators with problem settings $n_1 = n_2 = 64$, and $\sigma^2 = 10^{-10}$. The results are aggregated over 10 Monte Carlo runs at each specification of r, k, and m. An algorithm succeed to the recovery in the region below its phase transition curve

or r, and also as the decrease of m/n. We see that TMP has much border success recovery regions than P-BiG-AMP and SpaRCS.

4.6.4 Video Background and Foreground Separation

We now apply TMP to the video compressed sensing application to recover the background and foreground of multiple image frames of a video sequence from compressed measurements. The video sequence used in the experiments is of size $240 \times 320 \times 500$, consisting of some static backgrounds with a person moving in the foreground. We resize the sequence data into a matrix of size 76800×500. The compressive linear operator is chosen as the partial orthogonal linear operator given in (4.61).

We set measurement rate $m/n = 0.1$ and separate the background and foreground using TMP and SpaRCS. The recovery results are shown in Fig. 4.9. The first row of images are the original images with background and foreground. The second and the third rows of images are the recovered background and foreground, respectively, using the TMP algorithm. The fourth and the fifth rows of images are the background and foreground recovered by SpaRCS. Form the recovered images in Fig. 4.9, we see the estimated backgrounds and foregrounds using TMP are perceivably sharper than those using the SpaRCS algorithm.

4.7 Summary

In this chapter, we proposed a TMP framework for the design of message passing algorithms to tackle the compressed RPCA problem. The framework consists of three cascaded modules: one for linearly estimating matrices $\boldsymbol{L}$ and $\boldsymbol{S}$ based on

Fig. 4.9 TMP recovery results on a 240 × 320 × 500 video sequence. The video data is from [22]. The first row of images are ground truth for several frames. The second row of images are estimated backgrounds using TMP, the third row of images are estimated foregrounds using TMP, the fourth row of images are the estimated backgrounds using SpaRCS, and the fifth row of images are the estimated foregrounds using SpaRCS

the observed linear measurements, and the other two modules for refining the estimates by exploiting the low-rank property of $\boldsymbol{L}$ and the sparsity of $\boldsymbol{S}$. Various denoisers are considered for the low-rank matrix estimation module and the sparse matrix estimation module. We show that the proposed TMP algorithm significantly outperforms the state-of-the-art compressed RPCA algorithms such as SpaRCS, SPCP, and P-BiG-AMP in various system configurations. The application of video background and foreground separation validates the practicability of TMP on real data.

Appendix 1: Proof of Theorem 1

Since problems (4.7) and (4.8) are similar, it suffices to focus on problem (4.7). We have

$$\begin{aligned}
&\mathrm{E}[(S^{ext}_{A,i,j} - S_{i,j})(S^{pri}_{A,i,j} - S_{i,j})] \\
&= \mathrm{E}[S^{ext}_{A,i,j}(S^{pri}_{A,i,j} - S_{i,j})] && (4.62a)\\
&= \mathrm{E}[c_{S,i,j}(S^{post}_{A,i,j} - \alpha_{S,i,j}S^{pri}_{A,i,j})(S^{pri}_{A,i,j} - S_{i,j})] && (4.62b)\\
&= c_{S,i,j}\mathrm{E}[(S^{post}_{A,i,j} - \alpha_{S,i,j}S^{pri}_{A,i,j})(S^{pri}_{A,i,j} - S_{i,j})] = 0 && (4.62c)
\end{aligned}$$

where (4.62a) follows from $\mathrm{E}[(S^{pri}_{A,i,j} - S_{i,j})S_{i,j}] = 0$ and (4.62b) follows by substituting $S^{ext}_{A,i,j}$ with (4.6a). Then

$$\begin{aligned}
\alpha_{S,i,j} &= \frac{\mathrm{E}[S^{post}_{A,i,j}(S^{pri}_{A,i,j} - S_{i,j})]}{\mathrm{E}[S^{pri}_{A,i,j}(S^{pri}_{A,i,j} - S_{i,j})]} && (4.63a)\\
&= \frac{\mathrm{E}[S^{post}_{A,i,j}N^{pri}_{A,S,i,j}]}{\mathrm{E}[(S_{i,j} + N^{pri}_{A,S,i,j})N^{pri}_{A,S,i,j}]} && (4.63b)\\
&= \frac{v^{post}_{A,S,i,j}}{v^{pri}_{A,S,i,j}} && (4.63c)
\end{aligned}$$

where (4.63a) follows from (4.62c), (4.63b) follows from (4.5a), the denominator of (4.63c) follows from (4.5a), and the numerator of (4.63c) follows from (4.3b) together with some straightforward manipulations.

We now minimize the objective function of problem (4.8b) over $c_{S,i,j}$. Note that

$$\mathrm{E}[(S^{ext}_{A,i,j} - S_{i,j})^2] = \mathrm{E}[(c_{S,i,j}(S^{post}_{A,i,j} - \alpha_{S,i,j}S^{pri}_{A,i,j}) - S_{i,j})^2] \tag{4.64}$$

with $\alpha_{S,i,j}$ given by (4.63c). To minimize (4.64), the optimal $c_{S,i,j}$ is given by

$$c_{S,i,j} = \frac{\mathrm{E}[(S^{post}_{A,i,j} - \alpha_{S,i,j}S^{pri}_{A,i,j})S_{i,j}]}{\mathrm{E}[(S^{post}_{A,i,j} - \alpha_{S,i,j}S^{pri}_{A,i,j})^2]}. \tag{4.65}$$

By noting $\mathrm{E}[(S^{post}_{A,i,j} - S_{i,j})S^{post}_{A,i,j}] = 0$ and $\mathrm{E}[S^{post}_{A,i,j}N^{pri}_{A,S,i,j}] = v^{post}_{A,S,i,j}$, we obtain

$$c_{S,i,j} = \frac{v_{A,S}^{pri}}{v_{A,S}^{pri} - v_{A,S,i,j}^{post}}. \tag{4.66}$$

The MSE of extrinsic estimate $S_{A,i,j}^{ext}$ is given by

$$\mathrm{E}[(S_{A,i,j}^{ext} - S_{i,j})^2]$$

$$= \mathrm{E}[(c_{S,i,j}(S_{A,i,j}^{post} - \alpha_{S,i,j} S_{A,i,j}^{pri}) - S_{i,j})^2] \tag{4.67a}$$

$$= \mathrm{E}\left[\left(\frac{\mathrm{E}[(S_{A,i,j}^{post} - \alpha_{S,i,j} S_{A,i,j}^{pri}) S_{i,j}]}{\mathrm{E}[(S_{A,i,j}^{post} - \alpha_{L,i,j} S_{A,i,j}^{pri})^2]}(S_{A,i,j}^{post} - \alpha_{S,i,j} S_{A,i,j}^{pri}) - S_{i,j}\right)^2\right] \tag{4.67b}$$

$$= \mathrm{E}[(S_{i,j})^2] - \frac{\mathrm{E}[(S_{A,i,j}^{post} - \alpha_{S,i,j} S_{A,i,j}^{pri}) S_{i,j}]^2}{\mathrm{E}[(S_{A,i,j}^{post} - \alpha_{S,i,j} S_{A,i,j}^{pri})^2]} \tag{4.67c}$$

$$= \mathrm{E}[(S_{i,j})^2] - \frac{v_{A,S}^{pri}}{v_{A,S}^{pri} - v_{A,S,i,j}^{post}} \mathrm{E}[(S_{A,i,j}^{post} - \alpha_{S,i,j} S_{A,i,j}^{pri}) S_{i,j}] \tag{4.67d}$$

$$= \mathrm{E}[(S_{i,j})^2] - \frac{v_{A,S}^{pri}}{v_{A,S}^{pri} - v_{A,S,i,j}^{post}} (\mathrm{E}[(S_{i,j})^2] - v_{A,S,i,j}^{post} - \alpha_{S,i,j} \mathrm{E}[(S_{i,j})^2]) \tag{4.67e}$$

$$= \mathrm{E}[(S_{i,j})^2] - \frac{v_{A,S}^{pri}}{v_{A,S}^{pri} - v_{A,S,i,j}^{post}} \left(\mathrm{E}[(S_{i,j})^2] - v_{A,S,i,j}^{post} - \frac{v_{A,S,i,j}^{post}}{v_{A,S}^{pri}} \mathrm{E}[(S_{i,j})^2]\right) \tag{4.67f}$$

$$= \frac{v_{A,S}^{pri} v_{A,S,i,j}^{post}}{v_{A,S}^{pri} - v_{A,S,i,j}^{post}} \tag{4.67g}$$

where (4.67a) follows from substituting $S_{A,i,j}^{ext}$ by (4.6a), (4.67b) from substituting (4.65), (4.67d) from combining (4.65) and (4.66), (4.67e) from $\mathrm{E}[S_{A,i,j}^{post} S_{i,j}] = \mathrm{E}[S_{i,j}^2]$ and (4.5a), and (4.67f) from substituting (4.63).

Appendix 2: Proof of Theorem 2

The constraint in problem (4.18) is given by

$$\mathrm{E}\left[(S_{B,i,j}^{ext} - S_{i,j})(S_{B,i,j}^{pri} - S_{i,j})\right]$$

$$= \mathrm{E}\left[(S_{B,i,j}^{ext} - S_{i,j})N_{B,i,j}^{pri}\right] \tag{4.68a}$$

$$= \mathrm{E}\left[S_{B,i,j}^{ext} N_{B,i,j}^{pri}\right] \tag{4.68b}$$

$$= c_{S,i,j}\mathrm{E}\left[(S_{B,i,j}^{post} - \alpha_{S,i,j} S_{B,i,j}^{pri})N_{B,i,j}^{pri}\right] \tag{4.68c}$$

$$= c_{S,i,j}\mathrm{E}\left[S_{B,i,j}^{post} N_{B,i,j}^{pri}\right] - c_{S,i,j}\alpha_{S,i,j}\mathrm{E}\left[S_{B,i,j}^{pri} N_{B,i,j}^{pri}\right] \tag{4.68d}$$

$$= c_{S,i,j}\mathrm{E}\left[S_{B,i,j}^{post} N_{B,i,j}^{pri}\right] - c_{S,i,j}\alpha_{S,i,j} v_{B,S}^{pri} \tag{4.68e}$$

$$= c_{S,i,j}\mathrm{E}\left[\mathcal{D}_{S,i,j}(\boldsymbol{S} + \boldsymbol{N}_B^{pri}; v_{B,S}^{pri})N_{B,i,j}^{pri}\right] - c_{S,i,j}\alpha_{S,i,j} v_{B,S}^{pri} \tag{4.68f}$$

$$= c_{S,i,j} v_{B,S,i,j}^{pri}\mathrm{E}\left[\frac{\partial \mathcal{D}_{S,i,j}(\boldsymbol{S} + \boldsymbol{N}_B^{pri}; v_{B,S}^{pri})}{\partial S_{B,i,j}^{pri}}\right] - c_{S,i,j}\alpha_{S,i,j} v_{B,S}^{pri} = 0 \tag{4.68g}$$

where (4.68a), (4.68b), and (4.68e) follow from Assumption 4.1, (4.68c) from (4.17), (4.68f) from (4.16), and (4.68g) from Stein's lemma [14]. From (4.68), we obtain

$$\alpha_{S,i,j} = \mathrm{E}\left[\frac{\partial \mathcal{D}_{S,i,j}(\boldsymbol{S}_B^{pri}; v_{B,S}^{pri})}{\partial S_{B,i,j}^{pri}}\right]. \tag{4.69}$$

Also note that the MSE of the extrinsic output of Module B is given by

$$\mathrm{E}\left[(S_{B,i,j}^{ext} - S_{i,j})^2\right]$$

$$= \mathrm{E}\left[(S_{B,i,j}^{ext} - S_{B,i,j}^{pri})^2\right] - \mathrm{E}\left[(S_{B,i,j}^{pri} - S_{i,j})^2\right] + \mathrm{E}\left[(S_{B,i,j}^{ext} - S_{i,j})(S_{B,i,j}^{pri} - S_{i,j})\right] \tag{4.70a}$$

$$= \mathrm{E}\left[(S_{B,i,j}^{ext} - S_{B,i,j}^{pri})^2\right] - \mathrm{E}\left[(S_{B,i,j}^{pri} - S_{i,j})^2\right] \tag{4.70b}$$

where (4.70b) follows from the constraint of problem (4.18).

From (4.70), the minimization of the MSE of the extrinsic output of Module B is equivalent to the minimization of $\mathrm{E}\left[(S_{B,i,j}^{ext} - S_{B,i,j}^{pri})^2\right]$ since $\mathrm{E}\left[(S_{B,i,j}^{pri} - S_{i,j})^2\right]$ is fixed. Then,

$$\mathrm{E}\left[(S_{B,i,j}^{ext} - S_{B,i,j}^{pri})^2\right]$$

$$= \mathrm{E}\left[(S_{B,i,j}^{ext} - S_{B,i,j}^{pri})^2\right] \tag{4.71a}$$

$$= \mathrm{E}\left[(c_{S,i,j}(S_{B,i,j}^{post} - \alpha_{S,i,j} S_{B,i,j}^{pri}) - S_{B,i,j}^{pri})^2\right] \tag{4.71b}$$

with optimal $c_{S,i,j}$ given by

$$c_{S,i,j} = \frac{\mathrm{E}\left[(S^{post}_{B,i,j} - \alpha_{S,i,j} S^{pri}_{B,i,j}) S^{pri}_{B,i,j}\right]}{\mathrm{E}\left[(S^{post}_{B,i,j} - \alpha_{S,i,j} S^{pri}_{B,i,j})^2\right]}. \tag{4.72}$$

References

1. A.E. Waters, A.C. Sankaranarayanan, R. Baraniuk, SpaRCS: Recovering low-rank and sparse matrices from compressive measurements, in *Proc. of Advances in Neural Information Processing Systems (NeurIPS)*, pp. 1089–1097, Cranada, Spain, Dec. 2011
2. K. Lee, Y. Bresler, ADMiRA: Atomic decomposition for minimum rank approximation. IEEE Trans. Inform. Theory **56**(9), 4402–4416 (2010)
3. D. Needell, J.A. Tropp, CoSaMP: Iterative signal recovery from incomplete and inaccurate samples. Appl. Comput. Harmon. Anal. **26**(3), 301–321 (2009)
4. W. Ha, R.F. Barber, Robust PCA with compressed data, in *Proc. of Advances in Neural Information Processing Systems (NeurIPS)*, pp. 1936–1944, Montreal, QC, Canda, Dec. 2015
5. J. Ma, X. Yuan, L. Ping, Turbo compressed sensing with partial DFT sensing matrix. IEEE Signal Process. Lett. **22**(2), 158–161 (2015)
6. Z. Xue, X. Yuan, J. Ma, Y. Ma, TARM: A turbo-type algorithm for affine rank minimization. IEEE Trans. Signal Process. **67**(22), 5730–5745 (2019)
7. Z. Xue, J. Ma, X. Yuan, Denoising-based turbo compressed sensing. IEEE Access **5**, 7193–7204 (2017)
8. D.L. Donoho, A. Maleki, A. Montanari, Message-passing algorithms for compressed sensing. Proc. Natl. Acad. Sci. (NAS) **106**(45), 18914–18919 (2009)
9. J.T. Parker, P. Schniter, V. Cevher, Bilinear generalized approximate message passing part I: Derivation. IEEE Trans. Signal Process. **62**(22), 5839–5853 (2014)
10. J.T. Parker, P. Schniter, Parametric bilinear generalized approximate message passing. IEEE J. Sel. Top. Signal Process. **10**(4), 795–808 (2016)
11. M. Bayati, A. Montanari, The dynamics of message passing on dense graphs, with applications to compressed sensing. IEEE Trans. Inf. Theory **57**(2), 764–785 (2011)
12. Y. Kabashima, F. Krzakala, M. Mézard, A. Sakata, L. Zdeborová, Phase transitions and sample complexity in Bayes-optimal matrix factorization. IEEE Trans. Inf. Theory **62**(7), 4228–4265 (2016)
13. P. Schniter, S. Rangan, A.K. Fletcher, Vector approximate message passing for the generalized linear model, in *Proc. of 50th Asilomar Conference on Signals, Systems and Computers (ASILOMAR)*, pp. 1525–1529, Aachen, Germany, June 2017
14. C.M. Stein, Estimation of the mean of a multivariate normal distribution. Ann. Stat. **9**(6), 1135–1151 (1981)
15. D.L. Donoho, De-noising by soft-thresholding. IEEE Trans. Inf. Theory **41**(3), 613–627 (1995)
16. T. Blu, F. Luisier, The SURE-LET approach to image denoising. IEEE Trans. Image Process. **16**(11), 2778–2786 (2007)
17. C. Guo, M.E. Davies, Near optimal compressed sensing without priors: Parametric SURE approximate message passing. IEEE Trans. Signal Process. **63**(8), 2130–2141 (2015)
18. E.J. Candes, C.A. Sing-Long, J.D. Trzasko, Unbiased risk estimates for singular value thresholding and spectral estimators. IEEE Trans. Signal Process. **61**(19), 4643–4657 (2013)
19. S. Rangan, P. Schniter, E. Riegler, A.K. Fletcher, V. Cevher, Fixed points of generalized approximate message passing with arbitrary matrices. IEEE Trans. Inf. Theory **62**(12), 7464–7474 (2016)

20. A. Aravkin, S. Becker, V. Cevher, P. Olsen, A variational approach to stable principal component pursuit, in *Proc. of Conference on Uncertainty in Artificial Intelligence (UAI)*, pp. 32-41, Quebec, Canda, July 2014
21. J. Wright, A. Ganesh, K. Min, Y. Ma, Compressive principal component pursuit. Inf. Inference J. IMA **2**(1), 32–68 (2013)
22. L.K. University, Dataset: Detection of moving objects, available at: http://limu.ait.kyushu-u.ac.jp/dataset/en/

Chapter 5
Conclusions and Future Work

5.1 Conclusions

In many applications, data signals contain intrinsic structures that can be exploited in data compression. We study three different types of structured signals, namely, sparse signals, low-rank signals and sparse plus low-rank signals. The compressive recovery of these structured signals is the focus of this book. We start from the turbo message passing principle that can be generally applied to problem of structured signal recovery. The main results of the book are summarized as follows.

In Chap. 2, we investigate the compressive sparse signal recovery. In particular, we first present brief introduction to the turbo compressed sensing algorithm and then the extension from turbo compressed sensing to denoising-based turbo compressed sensing (D-Turbo-CS). Extrinsic message updates are calculated based on the turbo message passing principle. The optimization of the denoiser parameters based on Stein's unbiased risk estimate is also discussed in this chapter. The per-iteration performance of the denoising-based turbo compressed sensing is given by the state evolution. We show in the numerical experiments that the proposed D-Turbo-CS outperforms AMP and EM-GM-AMP in sparse signal recovery and compressive image recovery.

In Chap. 3, we investigate the low-rank matrix recovery problem and the low-rank matrix completion problem. In particular, we first propose an algorithm framework named turbo-type affine rank minimization (TARM) based on D-Turbo-CS. We focus on a special family of operators termed ROIL operators for the low-rank matrix recovery problem. We show that with ROIL operators, the performance of the TARM algorithm can be accurately characterized by the state evolution. For the low-rank matrix completion problem, we present three different approaches to determine the parameters of the TARM algorithm. We show by numerical experiments that the TARM algorithm performs better than the counterpart algorithms.

X. Yuan and Z. Xue, *Turbo Message Passing Algorithms for Structured Signal Recovery*, SpringerBriefs in Computer Science,
https://doi.org/10.1007/978-3-030-54762-2_5

In Chap. 4, we investigate the compressed robust principal component analysis that recovers a low-rank matrix and a sparse matrix from the compressive measurement of their sum. In particular, we first describe an algorithm framework termed turbo-type message passing (TMP) that bears a similar structure to the D-Turbo-CS algorithm. The main difference is that the TMP algorithm combines the low-rank matrix estimation module with the sparse matrix estimation model. Extrinsic messages are passed between different modules. The parameters involved in the calculation of the extrinsic messages are determined by following the turbo message passing principle. We study various low-rank matrix denoisers and sparse matrix denoisers based on which extrinsic messages are calculated. We then analyze the convergence of the TMP algorithm. Numerical results demonstrate a significant performance advantage of the TMP algorithm over its counterparts in various settings.

5.2 Future Work

The works presented in this book offer possibilities for future extensions. Specifically, the following topics are of interest:

- Learning-based parameter selection:
  In this book, the proposed algorithms involve parameters to be optimized, such as descent step size, denoiser's shrinkage threshold, and linear combination coefficients of extrinsic messages. As the deep learning technology rises, it provides powerful parameter learning for various applications. We may consider optimizing parameters involved in the proposed algorithms by using the deep learning technology. A promising way is to unfold the proposed algorithms into feedforward neural networks and the involved parameters can be learned using back-propagation.
- Distributed inference problems:
  In this book, we only consider the scenario that compressive measurements of structured signals are collected in a centralized manner. In many applications such as the sensor network and federated learning systems, devices are deployed distributedly and measurements are taken separately in these devices where centralized processing of signals may be restricted due to the limited communication and bandwidth or device privacy. We may want to recover the structured signals from the separate measurements without the communication between devices and reveal the individual measurements to the center node. In this scenario, each local device updates the global signal estimate using their measurement and push their updates to the center node.

- Generalized linear problems:

  We mainly considered the standard linear inverse problems in this book. However, real applications often involve complex relations. A flexible generalization of the linear problem is the generalized linear problem that applies a link function on the linear measurement. For example, the phase retrieval problem that aims to recover the phase information when only the amplitude of the linear measurement is reserved and the binary classification problem that classifies according to the sign of the linear measurement.

Index

X. Yuan and Z. Xue, *Turbo Message Passing Algorithms for Structured Signal Recovery*, SpringerBriefs in Computer Science,
https://doi.org/10.1007/978-3-030-54762-2

Printed in the United States
By Bookmasters